DIVIDED WE FALL

The Story of the Paperworkers' Union and the Future of Labor

June 25, 2004

Bill -
Union Brother - whatever it takes
for as long as it takes -

Pete Ved

DIVIDED WE FALL

The Story of the Paperworkers' Union and the Future of Labor

by

Peter Kellman

THE APEX PRESS
NEW YORK

Published by The Apex Press.
The Apex Press is an imprint of the
Council on International and Public Affairs
777 United Nations Plaza, Suite 3C
New York, New York 10017
800.316.2739

CIPA 50
1954–2004

Library of Congress Cataloging-in-Publication Data

Kellman, Peter, 1946-
Divided we fall : the story of the Paperworkers' Union and the future of labor / Peter Kellman.
p. cm.
Includes bibliographical references.
ISBN 1-891843-23-0 (pbk. : alk. paper)
1. United Paperworkers International Union--History. 2. Strikes and lockouts--Paper industry--Maine--Jay--History. I. Title

HD6515.P33K45 2003
331.88'176--dc22

2003057899

Cover and interior design by M.E. McCourt.
Cover cartoon by Will H. Call, from *Paper Maker's Journal*, August 1908.
Typeset and printed in the United States of America.

The author is a member of UAW Local 1981.

® GCIU 77P

TABLE OF CONTENTS

ACKNOWLEDGMENTS vii

INTRODUCTION 1

CHAPTER 1: The People's Republic of Jay 3

CHAPTER 2: Corporations and the Common People 17

CHAPTER 3: Unions in the Paper Mills: 1884 to 1908 25

CHAPTER 4: Labor in the Early Twentieth Century: A Prelude to the Strike of 1908 37

CHAPTER 5: A Great Lesson: The Strike of 1908 41

CHAPTER 6: The Great Lesson Applied: The Strike of 1910 49

CHAPTER 7: Consolidating Power: The Slippery Slope — 1910 to 1921 65

CHAPTER 8: The Strike of 1921: Aftermath and Lessons 77

CHAPTER 9: The Corporate Republic 91

CHAPTER 10: The Creation of a Bureaucrats' Union: Organizing Business Leaders 95

CHAPTER 11: National Labor Relations Act 113

CHAPTER 12: Mine Workers and Paper Workers 127

CHAPTER 13: Coming Full Circle: Militancy Is Not Enough 131

CHAPTER 14: So What Is to Be Done? 145

CONCLUSION 167

APPENDIX 170

- Paperworkers' Chronology 170
- Works Cited 177
- Laws Cited 180
- Poems and Songs Cited 181
- Recommended Historical References for Unions in the Paper Mills 181

INDEX 183

Note:

Footnotes in this book are used as explanatory notes.

Endnotes are used for the purpose of documenting sources; the full citations are found in "Works Cited."

ACKNOWLEDGMENTS

Everyone from Locals 14 and 246, especially Bill Meserve, Dennis Couture and Robbi Lucarellie. The members of the Jay-Livermore Falls Working Class History project, with special thanks to Ray Pineau and Brent Gay. And also to Ray (from Jay) Pineau, whose country and union let him down, but he never let me down, not in ten years, not ever. Thanks.

Ed Bruno, Gus Carbonella, Jack Getman, Richard Grossman, Pat Herrick, Michael Hillard, John Hinshaw, Leah Levesque, Staughton Lynd, Sharon Markham, Ward Morehouse, Jane Anne Morris, Joe Pilati, Jim Pope, Virginia Rasmussen, Ray Rogers, Rebekah Yonan and Mary Zepernick for their comments, suggestions, editing, patience, love and advice. Jack Getman who wrote the "Betrayal of Local 14" and in so doing made it possible for me to write about the history of the paperworkers rather than the strike of 1987/88. Richard Grossman and David Dembo for editing the last draft.

The Fund for Labor Relations, Grant Abert and especially the Program on Corporations, Law & Democracy for spiritual and financial support at crucial times during this project.

To everyone else who puts up with me and my endless questions. And thanks to my wife Rebekah Yonan for her love, encouragement and support.

Finally, Charles Scontras, Maine Labor Historian, who banged on a lectern one summer's day and said it was a struggle for workers just to get toilets in the workplace! Before that they had to go behind a column or hold it until they got home. Thanks, Charlie, for teaching so many of us that nothing is given to us. We have to fight for everything. We even have to fight to keep the knowledge of our own struggles alive.

INTRODUCTION

In 1986 managers of the largest paper companies in the United States held a secret meeting in Atlanta. According to President Wayne Glenn of the United Paperworkers International Union (UPIU),

> There were six companies there. And they decided to go for concessionary bargaining. IP was the leader . . . and, you see, his people (John Georges — CEO of International Paper Company) denied there was ever a damn meeting. They lied about it, and lied about it, lied about it. Then I had a meeting with him in New York, me and a couple of vice presidents, and sat down with him, in front of the other people, and I faced him with it, and I told him I knew about the meeting and I knew who was there, and he admitted it, in front of the people that had been lying all the time and saying it never happened.[1]

The union responded to the paper industry's coordinated plan with a plan of its own. On March 20, 1987 the United Paperworkers International Union told the International Paper Company it would utilize Coordinated Bargaining tactics against the company if it refused to back off its concessionary demands. In turn, the next day, the company locked out 1,400 workers at its Mobile, Alabama mill. Three months later UPIU locals at International Paper Company mills in Jay, Maine; Depere, Wisconsin; and Lockhaven, Pennsylvania began a coordinated strike in support of their brothers and sisters locked out in Mobile and against the concessionary demands being made by the International Paper Company.

In 1987 I was working for the Maine AFL-CIO and was assigned to help the striking locals in Jay. My role in the strike was that of organizer, educator and strategist. My life for two years was inseparable from the fight.[2] When the strike was lost, I felt a responsibility to study and document the struggle, so I could explain the defeat to myself and to my brothers and sisters.

My first step was to catalogue and study the thousands of files, letters, newspaper articles and videotapes I had collected which documented the strike.[3] But I didn't learn anything I didn't already know.

My next step was to try and understand why the International Paper Company had provoked the strike. I read many books on corporate power written in the 1980s from both the corporate and critics' perspectives. I learned why corporations did what they did, but I didn't learn why *we* did what *we* did. Still no answers. I needed to look even deeper.

Finally I started to research the history of the paperworkers' struggles in the Jay area and the answers began to roll.

People and social movements learn from their mistakes and successes. But histories of workers' movements are rarely told and usually lost. By uncovering the history of the paperworkers' struggles in the first four decades of the twentieth century, I began to understand why the strike of 1987/88 was lost.

The paperworkers' unions were subjected to the same national corporate activity in the 1980s that the rest of organized labor was subjected to and that situation holds up historically throughout the twentieth century. And so this book, through the paperworkers, focuses on the historical forces which created the institution we know today as organized labor. That institution shaped the strike of 1987/88 and determined its outcome.

In exploring this history of the paperworkers and organized labor, this book strives to answer three critical questions:

- What forces and events created today's institution of organized labor?
- What is it about the institution of organized labor that keeps labor from building a *working class movement*?
- What are the positive lessons from the past that workers can draw on to propel labor into the future?

Endnotes

1. Kellman Papers. Interview with Wayne Glenn by Jack Getman, September 18, 1990.

2. My role in the strike is tracked in Getman.

3. Those files and a computerized index are now housed at the University of Maine, Orono, Maine, Folger Library, Special Collections Department, Kellman Papers, Jay Strike Collection. The videotapes are in Bucksport, Maine, at the Northeast Historic Film Inc., Jay Strike Collection.

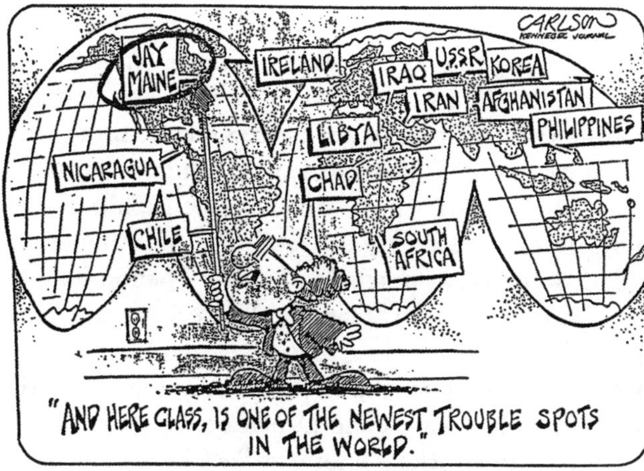

Carlson. *Kennebec Journal.* October 10, 1987.

1

THE PEOPLE'S REPUBLIC OF JAY

As was true of almost all important strikes during the 1980s and early 1990s, it was the company, not the union, that sought concessions. International Paper (IP) came to the bargaining table with a long list of demands — to lower labor costs, reduce jobs, and gain more management flexibility. When the union refused and went on strike in protest, IP hired workers to permanently replace the strikers. Local 14 fought an imaginative and effective battle, similar to strikes in the 1930s, marked by enthusiastic rallies, effective alliances, and grass-roots activism. [1]

It was hard for someone who lived in Maine during 1987/88 not to have heard about the Paperworkers' strike in Jay against International Paper Company, the dominant company in the paper industry. Almost every day for two years, strike-related stories appeared in the newspapers, and were aired on radio and television. The Maine media voted the strike the number two story in 1987, the number one story in 1988 and, 12 years after the strike ended, the state's largest newspaper wrote that the strike was one of the top events of the twentieth century in Maine and the only labor event to make their list. On the day the strike was called off, the largest television station in the state extended its nightly news show 30 minutes to cover the event.*

* Almost 1,000 articles on the strike appeared in the *Lewiston Sun*; the *Boston Globe* ran 45 articles in the paper and a feature story in its magazine section; and an article about the strike appeared in *The New York Times Magazine.*

One has to ask: what sequence of events awoke the sleeping labor giant in this small town located along the Androscoggin River in the foothills of western Maine? It started in 1986 when managers of major paper corporations met and agreed to a united front to get concessions from the unions. The first resistance to concessions demanded by the International Paper Company occurred in March 1987 when workers at the Mobile, Alabama mill rejected the company's concessionary proposal. The company responded by locking out the 1,400 union workers in Mobile.* Then in June 1987 workers at the company's mills in Depere, Wisconsin; Lockhaven, Pennsylvania; and Jay, Maine, responding to concessionary contract proposals and in solidarity with the workers in Mobile, struck the International Paper Company. Workers in Lockhaven and Depere set up picket lines but because they were involved in a lockout, not a strike, the Mobile workers just stayed home or found other jobs. The activity in Depere and Lockhaven revolved around getting union members to picket and convincing those contemplating crossing the line, not to.

In contrast to the picket lines in Lockhaven and Depere, Jay workers launched an offensive on many fronts.

- They organized mass demonstrations, issued press releases and held press conferences. Strikers leafleted at plant gates around the state and spoke at every forum available to them from the Chamber of Commerce to high school classes and local union meetings.
- The highlight of all the strike-related activity occurred every Wednesday night when 1,000 or more union members, their families, the public, and supporters from around the world gathered for a mass meeting held in the Jay Community Center.
- United Paperworkers International Union (UPIU) Local 14 in Jay also maintained a 24-hour-a-day picket line at the mill, a clothing bank, a job bank, a first-of-its-kind, drive-through food bank and formed a band to play at union functions.
- Union members were elected to the town's Board of Selectmen and the Maine Legislature. They increased the mill's tax base assessment in Jay by $100 million and denied the company a $4.5 million development bond.
- The Local introduced anti-strikebreaker legislation at the town meeting, the state Legislature and the U.S. Congress.

---

* After a contract expires and the negotiations come to a standstill the company can legally "lock out" union employees while temporary replacement workers are hired to fill union jobs. However, unlike in a strike where the company can hire permanent replacement workers once the strike begins, in a lockout if a new contract is signed the union workers return to their old jobs. In either case, union people refer to replacement workers as "scabs."

- Local 14 also promoted a precedent-setting local environmental ordinance which gave the town of 5,000 people the same authority as the Environmental Protection Agency (EPA) to issue licenses for land, water and air pollution, to monitor and enforce the law and fine the company for violations.
- Caravans of strikers crisscrossed New England and spoke at union meetings and high school and university classes. They leafleted at plant gates and on city streets. They gained media attention so favorable that the company sent its own team to speak in the same cities and towns visited by the union caravan and bought TV and newspaper advertisements to counter the union's positive image and growing public presence.

In Maine, Local 14 organized and led marches and demonstrations the likes of which had not been seen in Maine since the CIO organizing drives and strikes of the 1930s and 1940s. Local 14 was putting the *movement* back in the Labor Movement.*

As the strike in Jay picked up steam, workers and managers in the paper industry across North America followed it with a passion through the mass media and in union and industry publications. Many observers soon came to the conclusion that the future of labor relations in the industry hinged on the outcome of this one strike. In fact, as the power of the strike was felt throughout the industry, some paper companies put negotiations with their unions on hold. Others backed away from the concessions they were demanding, saying, "We don't want another Jay in any of *our* mill towns."** It appeared, for awhile, that the labor movement among paper workers had been reborn.

It was that breeze, created by a silent affirmation of dignity, that put the managers of the International Paper Company on notice: they were in for a long fight.

Local 14 was having an impact on the industry and the people who lived in paper mill towns around the country. What began as a strike to protect wages, benefits and working conditions was transformed into a labor movement that took power in the Town of Jay and exerted tremendous influence at the Maine Legislature. The strike had an impact on the Democratic presidential primary in 1988 and reached

* Local 14 of the United Paperworkers International Union (UPIU) represented 1,150 workers and Local 246 of the International Brotherhood of Firemen and Oilers represented 110 workers at the International Paper Company in Jay. The two locals combined forces during the strike, so for simplicity's sake when Local 14 is referred to it is implied that Local 246 was also involved. These two locals were decertified in 1992. Office workers at the mill were and still are represented by Local 442 of the Office and Professional Employees International Union (OPEIU), which did not strike but supported the two striking locals.

** Union negotiators at Scott Paper Company, Fraser Paper Company and James River Paper Company reported these company sentiments to mass meetings in Jay. UPIU Region 1 Director Jimmy Dinardo held a press conference to publicize this extraordinary activity on September 5, 1987 at the state capital in Augusta.

Images of the Strike

Many images characterize the strike for me, but the following particularly stand out:

- Every Wednesday night for 60 weeks more than 1,000 people assembled in the Jay Community Center. At the end of every meeting they stood, held raised hands, and with tears running down many faces sang:

> *When the union's inspiration through the workers' blood shall run.*
>
> *There can be no power greater anywhere beneath the sun. Yet what force on earth is weaker than the feeble strength of one?*
>
> *But the union makes us strong.*
>
> *Solidarity forever, Solidarity forever, for the union makes us strong.*

- Standing at the main gate of the mill on August 1, 1987 during a demonstration, I looked back two miles and saw 10,000 people heading up the road.

- One night at the weekly mass meeting in Jay, Lucien Deschaine the International Representative of the UPIU from Madawaska, Maine, brought greetings and thanks to the people of Jay from the people of Madawaska. The Paperworkers' local union in Madawaska had just negotiated a contract with Fraser Paper Inc. in which the union made no concessions. The people of Madawaska wanted to thank the people of Jay because they knew that the success of their negotiations was a direct result of the struggle being waged by the people in Jay.

- Pro-union families hung blue ribbons in their yards during the strike to let union people know that whatever the problem, they could stop at that house and find a friend. So if you had a flat, needed directions, a place to sleep, a meal or someone to talk to, you could just stop at a house with a blue ribbon flying.

- One night at a mass meeting a "super-scab" (a local union member who crossed the line) stood before more than 1,000 people and asked his parents, and his community to forgive him. He had crossed the line and worked in the mill for three days before realizing he had done wrong. I doubt if there was a dry eye in the hall that night as the community voted to let him back in the union.

- One night in August 1987 at a Jay Town Meeting, the citizenry came together to vote on three strike-related ordinances. When the first vote was taken, I was standing to one side of the hall and could feel the air move, like a refreshing breeze on a hot summer day, as the hands of those voting in the affirmative were raised. It was that breeze, created by a silent affirmation of dignity, that put the managers of the International Paper Company on notice: they were in for a long fight.

- The union hall was open and staffed 24 hours a day, seven days a week for the entire 16 months of the strike, except for one day. On February 5, 1988 an accident at the mill resulted in the release of 100,000 gallons of chlorine dioxide, a deadly gas, forcing a mass evacuation of the town. Had the temperature been 20 degrees warmer that February day, it is estimated that at least 3,000 people would have died.

- The last image, and one that speaks directly to the power generated by the strike, took place at the state capitol in Augusta. A delegation of strikers walking down the hall was spotted by a group of legislators who fell silent when they saw the strikers. Finally, one legislator spoke up and to all in hearing range exclaimed, "Here comes the delegation from the People's Republic of Jay."*

---

* Fourteen years after the strike, the State Representatives from Jay and the neighboring town of Livermore Falls are both former strikers.

Congress with the introduction of a bill to outlaw the use of permanent replacements.*

Three Sides

There were three institutions involved in this struggle: the International Paper Company, the United Paperworkers International Union (UPIU) and Local 14. At the time of the strike, International Paper Company was the largest paper company in the world and the largest private landowner in the country. It operated 26 primary pulp and paper mills and more than 100 secondary mills employing 50,000 people.** In 1986 it declared profits of $303 million; in 1987, $407 million; and in 1988, $754 million. The UPIU had more than 200,000 members, most of whom were employed in the pulp and paper industry. Local 14 represented 1,250 workers employed at the Jay mill. Each of the three institutions had something to win and lose. Each had a plan and its own degree of commitment to that plan.

The Company

In the fall of 1986 International Paper Company President Paul O'Neil, who 16 years later served in President George W. Bush's Cabinet as Secretary of the Treasury, was dispatched to western Maine to spread the word that the company wanted concessions and was serious about cutting wages in order to boost profits.*** O'Neil laid out the company's demands:

- End double-time pay for Sunday work (production workers were required to work 39 Sundays a year and the pay cut averaged out to $5,000 per worker per year).
- Eliminate the only remaining holiday the mill shut down, Christmas.
- Cut 350 union jobs by subcontracting all maintenance work.
- Phase out 170 jobs through a program the company called Project Productivity.

The company wanted the Town of Jay to cut the mill's property tax rate and to top it off, the company wanted from the state fewer environmental regulations and a decrease in Workers' Compensation rates.[2]

---

* Local 14's support of Jesse Jackson's candidacy helped him finish first in the Maine Democratic presidential caucuses, destroying the myth that Jackson didn't have credibility with white voters. Local 14 initiated the introduction of federal legislation to outlaw the use of permanent replacements during strikes at a time when the AFL-CIO had planned to keep the issue off its legislative agenda. After Local 14 became involved, the AFL-CIO changed its course (Getman 102-3).

** According to *Moody's Industrial Manual* (1987, 489), the company owned 7 million acres of woodland, operated 26 pulp and paper mills, 62 packaging plants, 4 envelope companies, 5 paper product merchandising companies and 20 wood product facilities in the U.S. It also owned 19 facilities in Europe, Latin America, and Asia.

*** O'Neil left the International Paper Company in 1987 to head the Alcoa Corporation.

The company's plan was simple. First, make a contract offer the union had no choice but to reject.* Then when the strike started, permanently replace the strikers as soon as possible, hoping to scare the membership into crossing the picket lines or voting to accept the contract. In the process, replaced workers would lose their jobs and the union, if there was one left, would be demoralized for years.**

The company had many things going its way. The tenor of the times was certainly against workers taking on a major corporation. In 1981, President Ronald Reagan had fired the nation's air traffic controllers for striking. Most of the strikes by unions that stood up to the concessionary corporate onslaught which followed (like the 1983 Steelworkers strike against the Phelps Dodge Company in the copper industry; the 1985 United Food and Commercial Workers strike against the Hormel Company in the food industry; or the 1986 United Paperworkers strike against Boise Cascade Corporation in Rumford, Maine) ended in defeat for the unions. In these strikes and in many more, the unions involved and the AFL-CIO were not able to mobilize the power of union members and their allies necessary to win.

The company's position also benefitted from the fact that at some of its other mills where contracts had not expired, the company could produce the same paper that was being produced in Jay. In addition, the company's vast financial resources could carry it through a long strike — if the strike was confined to a few mills. Finally, the company managers were committed to the concessions so they could get a 15 percent return on investment at all facilities, or so they claimed. And the union people be damned.

The National Union

About a year before the Jay workers struck, workers at the Boise Cascade Corporation mill in Rumford, Maine, 20 miles up the river from Jay, struck over many of the same issues that caused Jay workers to strike in 1987. Strikers had received $55 a week from the UPIU strike fund and not much else. Within three months, the corporation had permanently replaced 347 strikers and the union threw in the towel. It appeared that the national union's strategy was based on making the company lose money, primarily by keeping the skilled operators from crossing the picket lines. Such a strat-

* The contract the company proposed eliminated close to half the union jobs in the mill. It is almost impossible for people to vote in favor of eliminating their jobs. Thus, the company's proposal almost guaranteed the union's rejection and a strike.

** One month into the strike the company had permanently replaced 500 workers and by the end of the summer the entire work force had been replaced. For details on the rate of replacement, see the chronology of the strike in the Appendix.

egy may have worked in the 1950s and even as late as the 1970s, but it didn't have a chance against the corporations that dominated the paper industry in the 1980s. Paper companies like International Paper and Boise Cascade had abandoned the big labor/big corporation detente established in the United States after World War II and replaced it with "a take this concessionary-contract-in-times-of-record-profits or we will permanently replace you if you strike" — slash and burn labor policy. This policy was given a stamp of approval by President Reagan when he refused to negotiate in good faith with the air traffic controllers in 1980 and then fired them when they went on strike.

Local 14 and the UPIU

Workers in Jay had been very supportive of the Rumford strikers, sending money and helping out on the picket line. The Rumford experience made the Jay workers painfully aware that just a picket line would not do them much good when their contract ran out the next year.* So they started early to plan a strategy and assemble the resources necessary to win. Even as the strike in Rumford went on in the summer and fall of 1986, Local 14 began to mobilize its membership. In December they proposed to the company that negotiations get started early.

By offering to start negotiations early and to take a pay cut, the local took the offensive in appealing to the public for support. In its first offer, the local proposed a one-year extension of the contract; the second proposal was for a two-year extension. With inflation running at 4 percent a year the local was therefore proposing an 8 percent cut in real wages.[3] The company told the union on both occasions it was not interested in negotiating until just prior to the expiration of the contract.

The third part of the pre-strike program for the local was to put pressure on the UPIU to come up with a union-wide strategy to take on the International Paper Company. Local 14 and the other locals that had contracts coming up in June 1987 precipitated a series of meetings with the leadership of the UPIU. Out of these meetings came the UPIU's strategy.

The UPIU's Plan

Strike

- Under the plan, the locals at all International Paper Company mills in the United States were to strike as their contracts expired.

* The Rumford workers never forgot the Jay workers. During the Jay strike, the Rumford local presented an $11,000 check to Local 14 at a mass meeting every month.

Voting pool — "Coordinated Bargaining"*

- A voting pool was formed whereby locals that joined would have their votes pooled with the votes of the other members of the pool when it came to voting on company contract offers. Thus a contract proposal would be accepted or rejected by the entire pool at once and no one local would ever have to stand alone. The theory was, as contracts expired, the size of the pool would grow and so would its power. This concept was called "Coordinated Bargaining" by the UPIU.

No contracts signed

- The president of the UPIU, Wayne Glenn, promised he would not sign contracts at any International Paper Company locations, whether or not they were in the pool, until the company met the conditions set down by those in the pool.

Corporate Campaign**

- The UPIU leadership agreed to start what they called a "corporate campaign" against the company.

Strike pay

- The UPIU strike fund paid $55 a week to those members on strike. A month into the strike, the UPIU announced strikers would get $200 a week in strike pay. The plan was to get most of the over 200,000 members of the UPIU to donate $10 per member per month to the effort. The UPIU announced the plan at a press conference, raising the expectations of the strikers. However, the UPIU never came close to its goal of raising strike benefits to $200 a week because it never made an all-out attempt to involve its non-striking membership in the cause.

Local 14 Prepares the Soil

Local 14 was very active in the summer of 1986 supporting the Rumford strikers. The local also became involved in the 1986 state legislative races, using the opportunity to build an organization that combined in-plant communications with political action. It elected a very pro-union UPIU member to the Legislature while laying the groundwork for a strike organization.

In the spring of 1987, the local used the communications network built in the previous fall's political campaign to circulate a questionnaire among the membership. The questionnaire was used to educate the membership about the upcoming negotiations and asked members to get involved in committees that were being formed in anticipation of a possible strike. Work started soon after on building and training media and food bank committees. In

---

* Coordinated Bargaining can take place without strikes. The UPIU claims to have used the tactic successfully in negotiations with the Champion Paper Company in 1986.

** For a description of the Corporate Campaign see "Labor's New Sit Down" in Chapter 13.

May 1987 the local organized a press conference in Augusta and its first community mass meeting in Jay. The key speaker at both events was Larry Funk, the chair of the union's negotiating committee for the International Paper Company workers locked out in Mobile, Alabama.

In preparation for a confrontation with the International Paper Company, Local 14 had lined up political support in the legislature, built an active organization among its own members, had in place media and food bank committees and developed strong links to other locals in the International Paper Company system. They had prepared the garden bed well.

The Strike Begins

In June 1987, workers struck International Paper Company mills in Depere, Lockhaven and Jay in response to the company's concessionary demands and the lock-out in Mobile. The strike in Jay began on June 16 — 85 years to the day that the forerunner to the Pulp & Sulphite Union signed its first contract with the International Paper Company. In that 1902 contract, workers won three shutdown holidays (Labor Day, Independence Day, and Christmas Day); Sundays off; and if the company wanted the workers to run the mill on Sunday, the union had to approve and the company had to pay double time. By 1987 workers were fighting to keep Christmas, the only remaining shutdown holiday, and double time on Sunday. But in 1987, the *idea*, never mind the *right*, of the union having to agree to Sunday operation had been erased from the historical memory of the Paperworkers Union.

As the strike got underway, Local 14 arranged to send delegations of strikers in campers, cars and pick-up trucks across the country to visit workers at other International Paper Company mills. The plan was for these delegations to greet workers at the plant gates, hold mass meetings and visit workers in their homes. After the plans for the "outreach project" had been formed and the request for volunteers made at the first weekly mass meeting, UPIU Representative Royal Roderick told Local 14 leaders that the UPIU would take care of that type of activity. While waiting for the UPIU to put its plan into action Local 14 set up a phone bank to call local leaders in the over 100 locals in the International Paper Company system every week to give them an update on the strike and get information about activities in their mills.* Finally in April of the following year the "outreach program" was imple-

---

* Local 14 had to generate its own complete list of the other locals because the UPIU only had a list of the locals it represented. There were many other locals in the International Paper Company system represented by other unions like the Electricians (IBEW), Machinists (IAM) and Office Workers (OPEIU). There was no organization specifically representing all the locals in the system. Interestingly, Local 14 got a complete list of all the unions in the system from the company (unbeknownst to the company, of course).

mented when Local 14 put 25 full time people on the road, cris-crossing the country, visiting workers in International Paper Company mill towns.

Exhilarating and Exasperating

The first summer of the strike was both exhilarating and exasperating. The weekly meetings, the mass demonstrations, the support pouring in from all over the country, the solidarity, all made for a climate of hope in the strike community. But there were clouds on the horizon: the corporate campaign the UPIU had promised went the way of the $200-a-week strike pay and its non-existent "outreach project."

And then there was Pine Bluff.

Pine Bluff and Hamilton

The striking locals assumed that the UPIU leadership would leave no stone unturned in bringing other locals into the pool and encouraging them to strike.* But that assumption was called into question in August when the Pine Bluff, Arkansas, contract was set to run out. Because the Pine Bluff mill made the same paper as the Jay mill, it is likely that the company would have caved if workers there had struck. UPIU President Wayne Glenn spoke at a mass meeting in Jay on August 12, 1987, and was asked from the floor, "What is going to happen in Pine Bluff ?" Glenn responded that Pine Bluff workers would vote on a contract and make their own decision. But he assured the mass meeting that no matter the outcome, he would not sign the contract. However that would not be enough. Everyone in the hall that night knew Pine Bluff was key to bringing an early end to the strike. If Pine Bluff went out the International Paper Company would lose many of its important customers. Also if Pine Bluff struck, it would encourage other locals to strike and join the pool.

Meanwhile, local leaders in Jay were getting mixed messages from and about Pine Bluff. The phone committee didn't get the sense that much was going on down there to involve people in the struggle. But Joe Bradshaw, regional Vice-President of the UPIU looking after the Pine Bluff local, assured the strike leaders that Pine Bluff would do the right thing — join the pool and strike.

Local union leaders in Jay, Lockhaven, Mobile and Depere, deeply concerned about the situation, sent a delegation to Pine Bluff to talk to the locals first

---

* UPIU locals can not legally call a strike on their own. The local must first vote to strike and then the UPIU can sanction the strike or not. A strike cannot be legal without being sanctioned by the UPIU. The process to call off a strike only involves the President of the UPIU (see works cited — Agee-Glenn). He could decide to require a vote of the membership or make the determination himself.

hand. When the delegates from Jay arrived, they got a cold reception from Tommy McFalls, the UPIU International Representative in charge of negotiations, and from the local union leadership. It appeared the fix was in. Pine Bluff voted to accept the contract.

In September the striking locals found out that another big mill in Hamilton, Ohio, got a one-year extension. Under International Paper/UPIU contracts, if neither party serves notice to the other that it wants to negotiate at least 60 days prior to the termination of the contract, then the contract remains in force without change for a year. The UPIU had until the end of July to serve notice that it wanted to terminate the contracts in Hamilton, Ohio and Corinth, New York. It didn't, but the company did in Corinth. The UPIU claims it just made a mistake and meant to serve notice at both locations. However, given that the strikes going on at the time were the national union's most important strikes since 1921 and that the most important obligation of the national union is to conduct contract negotiations, the union's excuse would not pass the straight-face test. It was clear by September 1987 that the leadership of the UPIU was undermining the very programs they had proposed.

But the Locals Fought On

By September 1987 the company had permanently replaced the entire union workforce in Jay, Lockhaven and Depere, leaving the striking locals no alternative but to fight on. They did, but the die had been cast. At times over the next year, it appeared that the locals had turned the corner and could see the light at the end of the tunnel. But as things turned out, it wasn't daylight they saw. It was the UPIU with a candle and the company with a search-and-destroy light. Finally, on October 10, 1988, the UPIU blew out the candle and called the strikes off without a vote of the membership. Shortly thereafter a striker wrote this letter:

Mark Well That in This Place You Stood

by Phil Edwards

Brothers and sisters, family members, friends of labor:

My father built bridges under fire in France in World War I and my brother died in a B-17 over England in World War II. I landed on a beach in North Korea in 1950, and, oh, the country and the government were behind us in those days. But, in 1987, in Jay, Maine, we had a war about corporate greed and the depredation of worker's rights and where then was the country and where then was the government? Did the Governor of the State or the President of the United States stand up to corporate America? They did, in fact, encourage the pillaging of workers' hard-earned benefits. Where was the United States Congress? Where was the national media during our struggle?

What does the International Executive Board think is the important issue here? Are they concerned that they will deplete their funds? I tell you, they should have been ready to sell everything they own to win this conflict. Are working people anywhere naive enough to think this is not their work?

To those of you who helped, and your numbers are legion, I give my thanks, my most earnest heartfelt thanks, but to those who did not, I would paraphrase John Donne: "Ask not for whom the bell tolls. It tolls for thee."

Sisters and brothers, I do not have the words to ease your pain on this sorry occasion. Each of you have given of your time, your energy, and your personal fortune to advance the cause of labor during this dark age. History will record that you, like the embattled farmers at Concord Bridge, fired a shot heard 'round the world. Do not despair that the battle was lost along another front. *Mark well that in this place, you stood*, shoulder to shoulder, faced the enemy, and did not flinch nor bow down.

I have known some of you for most of the thirty-odd years that I have worked at International Paper. I have found you to be honest tradesmen, in the Maine tradition. Since the strike began, I have come to realize how much I value your friendship. I have seen lambs become lions and sparrows become eagles. From your ranks have emerged salesmen and lawyers, organizers and orators.

You have been tempered into steel in the crucible of war, and you will never again be as you were before.

I could not allow this moment to pass without telling you how proud my wife and I are to have soldiered with you.[4]

So we now leave that strike to go back in time and trace the evolution of the paperworkers' unions, the times their members lived in, the struggles they waged and the forces they were subjected to.*

Endnotes

1. Getman xi.

2. Charles Coolidge. "Industrial Survival-squabbling Must Stop," *Lewiston Sun.* October 30, 1986.

3. According to the U.S. Department of Commerce's consumer price index, the increase was 3.6 percent in 1987 and 4.1 percent in 1988.

4. *Pain on Their Faces.* 6.

* For anyone interested in reading a detailed account of the strike of 1987-88 in Jay, I highly recommend Jack Getman's, *The Betrayal of Local 14.* For those interested in reading an account of what the strike meant to the participants, *Pain On Their Faces* is a book of essays written by Jay strikers. The intent of the writers is to leave for those who follow in their footsteps an account of what the struggle meant to the participants, unedited, in their own words.

CORPORATIONS AND THE COMMON PEOPLE 2

Nearly a century ago the large corporation, as we know it today, first appeared as a solution to the ills of a competitive marketplace. As the strategy of combination supplanted competition, corporations began to consolidate through various means, or simply to gobble up their competitors in rapid fashion. By 1904, the first great merger movement of corporate capitalism had utterly transformed the structure of the American economy and set in motion the corporate reconstruction of American society.[1]

Towards the end of the nineteenth and the beginning of the twentieth century, large corporations built industrial mills along the banks of Maine rivers, among them some of the first modern pulp and paper mills in the country. The Maine Commissioner of Industry and Labor Statistics reported in 1899, "This great [pulp and paper] industry has been almost entirely developed since the Civil War, and the greater part of it since 1880."[2] In 1890 in Maine there were 17 pulp and paper mills employing 1,568 people; in 1900 there were 35 mills employing 4,851 people and in 1909 there were 45 mills employing 8,647 people.[3]

One of corporations, the International Paper Company, was formed by the 1898 merger of 17 leading paper mills. Based in the northeast part of the United States, it was the largest producer of newsprint in the nation.[4] The corporation was chartered in New York State and included several mills in Jay and Livermore Falls, Maine. These mills had been the property of one of International Paper Company's founders, Hugh Chisholm.

"The Corporation," as the company referred to itself at the time, constructed a modern paper mill, the new Otis Mill, on the banks of the Androscoggin River in the towns of Jay and Livermore Falls.* To build this mill, it brought in 500 Italian contract laborers and housed them in a place called Italian Camp,[5] where:

> Penniless, largely illiterate, and completely ignorant of the new tongue of their adopted country, these were welcome prey for the employment agency. In return for a job, the agency received a per-

* In 1887 Chisholm started the Otis Falls Paper Company and built the first Otis Mill. *The Boston Herald* reported in 1889 that the Otis Mill was the largest mechanical wood pulp mill in the state (Sturtevant 123).

> centage of the workers' pay which during this period, averaged 17.5 cents per hour. In addition the agency had control of the commissary .. .. .. Compelled to buy food and articles of necessity from the commissary where exorbitant prices were charged, there was usually little, and more often nothing, for these people to show after a month of work.

There was nothing unusual about the conditions these Italian contract laborers lived and worked under; all over the country in mills, factories, mines and on the sea similar conditions existed.

The new mill was completed in 1908 and run by the International Paper Company until they sold it in 1978. The mill is still in operation today.

Paper Makers' Journal. September 1908.

PAPER TRUST

"When the International Paper Co. was formed, it was bitterly assailed ...by the American Newspaper Publishers' Association" and a statement was signed to that effect by the owners of 157 newspaper publishers in the United States.[6] The Business Manager of *The New York Times* testifying before the Industrial Commission in Congress on April 12, 1901, accused International Paper Company of being the "Paper Trust." "All the big profitable paper mills, with a few unimportant exceptions, were merged into the company, twenty-four mills in all, producing about 80 per cent of the American output at the time."[7]

Again, in 1904, the company was accused of being part of the Paper Trust by *The New York Times, The Philadelphia Ledger* and other members of the American Newspaper Publishers' Association.* Testifying in Congress for the Association, Mr. Seitz of *The New York World* newspaper said,

> Moreover we have reached the point where the trust is beginning to dominate opinion and edit newspapers. Several years ago, when I tried to negotiate with the Paper Trust, the selling agent brought to my office a written schedule of commissions under which they would permit themselves to do business, and one of these was the declaration, 'It is understood that the publisher of The World must print nothing hostile to the interests of the International Paper Company'.[8]

Labor played a major role in opposing the formation of trusts and monopolies or control by one or several companies over an entire industry. A federal law, the Interstate Commerce Act (1881), was supposedly passed to regulate railroad monopolies and the Sherman Act (1890) was enacted to try to end the control corporate business monopolies had over sectors of the U.S. economy. But in the end large corporations like the International Paper Company in the paper industry, the Standard Oil Company in the oil industry, and the Dupont Company in the munitions industry, ended up setting prices and wages despite the law.

On its face, the Interstate Commerce Commission (ICC) had been set up to protect the public from railroad monopolies, but President Grover Cleveland's U.S. Attorney General Richard Olney pacified business interests with the explanation that the ICC:

---

* A trust is a combination of firms or corporations for the purpose of reducing competition and controlling prices throughout a business or an industry.

> is, or can be made, of great use to the railroads. It satisfies the public clamor for government supervision .. .. Further, the older such a commission gets to be, the more inclined it will be to take the business and railroad view of things. It thus becomes a sort of barrier between the railroad corporations and the people.. .. .. The part of wisdom is not to destroy the Commission, but to utilize it.[9]

The Sherman Act was also used as a foil to pacify the demands of the populist and labor movements. It allowed politicians to say they were doing something about the trusts but in reality they did little. Jeffrey Lustig observes in his book *Corporate Liberalism*,

> The ubiquitous Olney also explained why the Sherman Anti-Trust Act was invoked only once during the period of rapid combination. 'You will have noted that the government has been defeated on the Trust question,' he wrote a friend after the famous sugar monopoly case of 1895. 'I always supposed that it would be and have taken the responsibility of not prosecuting a law I believe to be no good.[10]

Matt Wuerker

On the other hand, corporations played a major role in using government to keep workers' organizations weak and disorganized. For example, they got the courts to issue injunctions against pickets, strikes and boycotts.* John Mitchell, President of the United Mineworkers Union (1898 to 1907), wrote in 1903:

> By means of it [the injunction] trade unionists have been prohibited under severe penalties from doing what they had a legal right to do.... It is difficult to speak in measured tones or moderate language of the savagery and venom with which unions have been assailed by the

* A boycott is an exercise of free speech in which people urge other people not to do something, like not buy a product or not patronize a store because it sells a particular product. Not buying a product is called a primary boycott. Not buying anything from a store that sells the product is called a secondary boycott.

> injunction, and to the working classes, as to all fair-minded men, it seems little less than a crime to condone or tolerate it.[11]

Labor argued that the injunctions were violations of the 13th Amendment and responded by violating injunctions and passing laws aimed at protecting the working classes. But once again the courts intervened by jailing workers for violating injunctions and declaring most of the pro-labor laws unconstitutional because they violated the "liberty" of corporations to freely enter into contracts with employees.

Image of Workers

If your image of workers on strike during the latter part of the nineteenth century is that of men standing around a plant gate with picket signs, think again. Between 1881 and 1900 in the State of Maine, there were 172 strikes involving 254 establishments in which 16,335 workers participated, about one-third of whom were female.[12] Between 1891 and 1900, only half the strikes were called by labor organizations. Class consciousness was so strong that workers often called strikes without the organizational backing of an official union. Strikes were lost at a rate of 2 to 1. During this period lockouts were just as common as strikes. These facts would lead one to believe that unlike today, workers at the turn of the century were not as afraid as

M.B. Schnapper, *American Labor; A Pictorial Social History,* Wshington D.C.: Public Affairs Press, 1975.

workers are today to challenge the managers of the corporations they worked for.

Another factor that contributed to this lack of corporate subservience was the role workers played in the production process. Today we think of "trade secrets" only as corporate property, but at the turn of the century it was workers who held most of the trade secrets. Trade secrets were their property. Workers' trade secrets were part of a "culture of control" at the work site. Corporate managers, at the turn of the century, began to take control of the workplace by instituting a production process (Taylorism/Fordism) where each worker was involved with only one small part of the overall process and a supervisor class of workers was created to coordinate and control the work site and production process. The result was workers lost their trade secrets, became alienated from their work, and the "culture of control" moved out of the workers' hands and into the realm of the corporate managers.

WHERE DID THEY COME FROM? WHAT DID THEY THINK?

The people who worked in the factories and mills were mainly immigrants or children of immigrants.[13] In Livermore Falls and Jay, most of the mill workers came from Quebec, the British Isles and Italy. These working people debated where their society was going and they built organizations to move the society in the direction they advocated. They had hope, were focused, and were a force to be reckoned with. In 1902 *The Paper Makers' Journal*, the official organ of the United Brotherhood of Paper Makers,[14] published a speech titled *Conditions of Labor in this Country* given by Ex-Governor of Illinois John Altgeld to a labor gathering in Brooklyn:

> You ask what will this corrupt commercialism lead to? I believe it will lead to better things; I believe it is the muddy crystals out of which will yet come the beautiful colorings of a new civilization. Without intending it, the leaders of this commercialism are hastening events toward an economic and cultural condition, in which there will be a more equitable distribution of the fruits of labor. Morgan and Rockefeller are putting the question squarely to the American people, whether the monopolies shall own the people or whether the people shall own the monopolies.
>
> Put in the concrete form, there is only one answer, and that answer is coming from all parts of the land. It is already written on the sky — that the people must own the monopolies. The trusts, by consolidating the industries and proving that they could be run from a common center, have paved the way to public ownership, wherein every toiler will be part owner of the shop. With

> all monopolies once in the hands of the people the solution of the industrial problems will be easy.*

When we look with today's eyes and try to understand the struggles of workers in a Maine paper mill town of the early 1900s we see working people had *hope, direction and organization.* Ex-Governor Altgeld put it this way at the end of his speech to the Brooklyn workers:

> Finally let me say that a new order of thought is abroad in the land. While at the top the corporations are strangling justice and robbing the people; underneath there is growing and rapidly spreading a general demand for more equitable conditions. All classes are expressing discontent with wrong, and a condemnation of the degrading commercialism of our time. An entirely new literature which breathes the spirit of human brotherhood is filling the land. The pen of the age is on the side of humanity. Only the hired scribblers for the press and the capitalist magazines are on the other side.
>
> These higher ideals will in time shape destiny and crystalize into new institutions. They will give to the world a new age and a new civilization. A civilization in which the hand of toil shall not be robbed of the fruits of its own labor. The immediate future seems dark with much tribulation, but the hirelings who now torture labor and assassinate liberty will be buried in their own infamy. A new morning will dawn, radiant with the splendors of freedom, and the children of toil will come into their inheritance.

Endnotes

1. Bowman 2.

2. *Annual Report 1899* 32.

3. Scontras 47.

4. Claimed a company spokesperson in an advertisement that appeared in a special section of the *Portland Press Herald* [ME] March 16, 1998, p.11 titled "Maine Paper Expo 1998." However, sources vary on the number (16 to 24) of properties and mills that merged to form the company.

5. Sturtevant. "500 Italian ...." 94-95; "Penniless...." 174.

* As Governor of Illinois, John Altgeld pardoned the three 1886 Haymarket martyrs who had not been executed. For this act of conscience, he lost a bid for reelection and was hounded by the corporate press for the rest of his life. However, he was honored by labor people wherever he went.

6. Collier 67.

7. Norris 7.

8. *New York Times.* April 8, 1904.

9. Lustig 94.

10. Lustig 94.

11. Mitchell 324.

12. *Annual Report 1904* 176.

13. Gutman 255.

14. Before 1911 the title of the official publication was the *Paper & Pulp Makers' Journal.* In 1911 it became the *Paper Makers' Journal.* Also the apostrophe after the word Makers' comes and goes over the years in the title and in the name of the union.

3

Union Membership by Year in Unions Nationally and in the Paper Industry 1900-1908			
Year	**Paper Makers**	**Pulp & Sulphite**	**All Unions**
1900	400	(not in existence)	900,000
1902	4,100	2,500	1,500,000
1903	10,700	4,500	1,900,00
1907	3,100	4,500	2,000,000
1908	4,300	4,500	2,000,000

Note: Data for different unions are not strictly comparable—provided for trends only.

UNIONS IN THE PAPER MILLS 1884 to 1908

Monday (September 8, 1903) will go down in history as the first time Labor Day was ever observed at Livermore Falls by representatives of the various labor unions;.. .. .. It was estimated that from 1,500 to 2,000 were in line... At an early hour the people began to fairly pour in from surrounding towns, and long before the arrival of the big special from Rumford Falls and from Lewiston, on the P. & R.F. Railway, the streets were just crowded with people.. .. ...[1]

As we saw previously, those who were directing the companies were building their organizations to make a profit, by having mills built and forming trusts to control the markets in which they bought the raw materials, sold the finished product and set the price of labor. Meanwhile, the workers who supplied the labor and made the product were developing unions and political parties to protect their own interests.

Internal Problems

The first period of union organization in the paper mills, 1884 to 1908, was rife with internal problems. Workers formed and reformed various unions until 1906, when two dominant unions emerged. However, from 1906 to 1908 these two unions almost destroyed each other, and by the end of 1908 they barely existed.

The names of the unions and whom they represented during this period are confusing. Historians writing about the period and journalists writing during the period often confuse the names of the competing unions.* I am not about to set the record straight here for all time, but what follows is a condensed description of what happened. It is important to point out that many of the same workers in these quarreling unions were also involved in trying to create broad social change through the Knights of Labor, the Socialist Party, the Peoples Party, the Industrial Workers of the World (I.W.W.), the eight-hour day movement, the movement to end child labor and by linking the labor movement to the prohibition and suffrage movements.

The First Union

Worker organization in the modern pulp and paper mills began in 1884 when an association of highly skilled paper machine tenders called the Eagle Lodge was formed in Holyoke, Massachusetts. The Lodge was a social club that evolved into a union by taking concerted action to improve conditions at work. Around 1891 the members of the Eagle Lodge, realizing that they needed more than a social club to change the working conditions in the paper mills, placed an assessment on themselves and hired an organizer. The organizer traveled to other mills and had some success organizing local unions. In 1893 the American Federation of Labor (AFL) granted the Eagle Lodge and its sister organizations a charter. The new union took the name United Brotherhood of Paper Makers (UBPM). The jurisdiction of the new union included highly skilled machine tenders and beater engineers. Beater engineers supervised the pulp mill and machine tenders were in charge of running the paper machines.

Coupled with a depression that made it hard to organize workers anyway, organizing just the "boss workers" into a union did not go well for the UBPM. By 1897 they had only three locals so they asked the AFL to grant them broader jurisdiction. The AFL obliged, but the machine tenders were unhappy in a union with common workers, so in 1898 they formed the International Paper Machine Tenders Union (IPMTU). Meanwhile the semi-skilled workers joined the Federal Laborers Protective Union, which appeared in Jay (Chisholm) in 1901 as the Laborers Protective Union No. 9,555 with 350 local members.**

---

* For a good description of the unions prior to 1906 see Burns.

** The neighborhood surrounding the Otis Mill in Jay is called Chisholm after the founder of the International Paper Company, Hugh Chisholm. Maps often show Chisholm as a Town, often in the place of Jay, as do numerous government reports. However, there is no town or official political entity in the town of Jay called Chisholm.

The First Contract

According to John Burke, president (1917 to 1965) of the Pulp & Sulphite Workers Union, the year 1902 marked the beginning of collective bargaining between the union and the International Paper Company.[2] On June 16 of that year, the company and the Federal Laborers Protective Union, the forerunner of the Pulp & Sulphite, signed a one-year agreement covering workers in six mills, including the Otis Mill in Jay, Maine.* In that agreement, Labor Day, Christmas Day, Independence Day and all Sundays were shutdown holidays when the mills could not operate unless the union agreed first. If the mills did operate workers were paid double time.

The pulp mill workers who were members of the Federal Labor Union felt that they needed a separate and distinct organization of their own and so they held a convention in 1902. Out of the convention came a request from the pulp mill workers to the AFL for a charter. But the AFL refused and told them to join the United Brotherhood of Paper Makers (UBPM), which by now had changed its name to the International Brotherhood of Paper Makers, Pulp, Sulphite and Paper Mill Workers (IBPMPSPMW) in order to reflect its new jurisdiction.

The Final Separation

The two groups could not get along. They were still at the bottom of the working class solidarity learning curve. The breakup began when local unions of the IBPMPSPMW held separate meetings, one for paper makers and one for the pulp workers. Finally, in 1906 the pulp workers held their own national convention and formed the International Brotherhood of Pulp, Sulphite and Paper Mill Workers (IBPSPMW). The AFL refused to recognize the Pulp & Sulphite. The result was war between the Pulp & Sulphite and the Paper Makers. The Paper Makers went on strike against International Paper Company in 1908. The Pulp & Sulphite Union workers crossed the Paper Makers' picket lines and scabbed the Paper Makers' jobs. Both unions were almost destroyed in the process.

The AFL recognized the Pulp & Sulphite in 1909. Finally there were two unions in the paper industry recognized by the AFL: one representing highly skilled workers, the other representing everyone else. The one representing the highly skilled was commonly referred to as the Paper Makers and the

* The other five mills were in Fort Edward and Ticonderoga, New York; Franklin Falls and Berlin, New Hampshire; and Rumford Falls, Maine.

Getting Rid of Superfluous Ballast.

Paper Makers' Journal. 1906.

other was called the Pulp & Sulphite, which are the terms I will use to describe the two unions after 1906.*

The trouble between the two unions did not end because the AFL decided to recognize both of them as separate unions with different jurisdictions. They stopped fighting because they learned, in the disastrous strike of 1908, that when they fought with each other, workers lost and the International Paper Company won.

* There were and are many unions that represent workers in the paper industry including the Carpenters, Fireman and Oilers, Plumbers, Electricians and Machinists. This book will only deal with the two unions that have represented most of the organized workers, the Paper Makers and Pulp & Sulphite. In 1972 the two unions joined forces under the name of the United Paperworkers International Union (UPIU) and in 1999 the Oil, Chemical and Atomic Workers Union (OCAW) and the UPIU merged to form Paper, Allied-Industrial, Chemical and Energy International Union (PACE).

The local status of the unions in the paper industry before and after the strike of 1908 is reflected in the annual reports of the Maine Commissioner of Industrial and Labor Statistics which reported in 1908;

> East Livermore (Livermore Falls)
>
> International Brotherhood of Papermakers, Pulp, Sulphite and Papermill Workers, No. 11. Secretary, Martin Burk, Livermore Falls; date of organization, March, 1902; has trade agreement with employers which expires June 25, 1908; number of members 110; qualifications for membership, one years work on paper machine; initiation fee, $1.00; monthly dues, 65 cents; times of meetings, third Sunday in each month; no benefits; hours of labor, 8; minimum daily wages, $1.65; maximum, $3.75; average time employed, 12 months.
>
> ---
>
> International Brotherhood of Pulp, Sulphite and Papermill Workers, Union No. 18, Secretary, B. N. Tretheway, Box 11, Chisholm; date of organization, 1903; has trade agreement with employers which expires June 1, 1908; number of members, 200; qualifications for membership, must be a laborer in or about a paper mill other than the engine plant and paper machines; initiation fee, $1.00; monthly dues, .35 cents; times of meetings second Sunday in each month; no benefits; hours of labor, tour workers, 8; day workers, 9; minimum daily wages, $1.80; average time employed, 12 months.

In contrast to the 1908 report, the 1909 report shows the membership of the Pulp & Sulphite down to 60 members. The Paper Makers didn't report any members in Livermore Falls, and in the entire state they only reported 273.

"SCAB IT" *

The war between the two unions went much deeper than name changes and the loss of membership. The divisive nature of this rivalry leading up to the strike of 1908 is illustrated in a letter written on April 27, 1906 by the national president of the Pulp & Sulphite Union to a local member in Jay, B.H. Trethewey.

> Your letter of the 25th received and contents carefully noted. Please accept congratulations on the stand taken by your membership, they show the right spirit when they decide they would prefer to scab it, (than) to again becoming subordinate to the Paper-Makers.

---

* A scab is anyone who takes a job from a union worker in a labor dispute. The term always has very negative connotations. Management prefers the term "replacement worker."

You know inter-union problems are bad when union workers talk about scabbing other union workers' jobs. But when they are encouraged by the national president of the union things are *really* bad.

One can surmise that relations were not just difficult at work. These people lived in the same town and it is likely that these problems did not stop at the mill gate. They probably divided the community and caused great friction among families, neighbors and friends.

An article in the *Paper Makers' Journal* for February 1902 answers the question, "What is a Scab?"

> According to these unionists a scab is to his trade what a traitor is to his country; and, though both may be useful to one party in troublesome times, when peace returns they are detested alike by all; so, when help is needed a 'scab' is the last to contribute assistance, and the first to grab a benefit he never labored to secure. He cares only for himself, but he sees not beyond the extent of a day, and for monetary and worthless approbation would betray his friends, his family and his country.. .. .. He is an enemy to himself, to the present age, and to prosperity.[3]

Paper Makers' Journal. May 1907.

Careless Operators?

The conditions inside the paper mills were terrible not just between the workers, but for all workers. Fingers, arms and lives were routinely squashed by the paper machines and pulp grinders. The inspector for the Maine Commissioner of Industrial and Labor Statistics stated in his 1895 report on pulp and paper mills:

> While perfection in the construction of these mills seems to be the point aimed at, the percentage of accidents is unduly large. But we must take into consideration the carelessness of the operatives as well as the character of the machinery used, for there is no class of workmen, so far as your inspector has been able to observe, who are so unmindful of the dangers from machinery as those working in these mills, the reason for which is not easily explained.[4]

Keep in mind that at this time the people who worked around the paper machines, in what was a very wet environment, worked in their bare feet for long hours on swing shifts in brutal heat. But the inspector must have missed that.

A letter written to a Wisconsin newspaper during the strike of 1902 lays out the conditions paper workers labored under:

> How many hours constitutes a fair, honest day's work? Is it eleven hours by day one week and thirteen by night the following week, in a heated and damp atmosphere, and standing on wet floors? Thirty years ago men started the paper industry in the Valley. Today they are millionaires. And yet these men refuse Sunday rest to the midnight toiler. And during all this controversy, no word is spoken for the poor young working girls who sit and cut paper all night long. And why? So printers may buy one-sixth to one-eighth of a cent per pound cheaper, and build mammoth structures for the great city dailies. There is not a dray horse in the Valley working 78 to 90 hours a week.....How many men over 50 years of age will you find running a machine today? And why? Because they have always had to work such unreasonably long hours that they cannot stand the strain and are broken down men today.....The State and federal Governments have established the eight hour day, and yet the paper makers are criticized for asking that 13 hours be given from a 78 hour week.[5]

In 1909 the Commissioner of Industrial and Labor Statistics was required by the Maine Legislature, for the first time, to investigate and report on industrial accidents. He reported that year:

> For some reason a great many accidents that happen in our industrial plants never reached the papers and consequently our report is incomplete to this extent. The list of accidents here presented numbers 338 covering the period between April 1 and December 1, 1909. Forty-one of them resulted fatally and most of the rest were of a serious nature.[6]

FORTY-ONE DEATHS in eight months. That is more than ONE DEATH A WEEK!!! And these workers didn't die of heart attacks.

> June 5, 1909 — employed as a second hand on a paper machine.. .. .. was caught between the first dryer roll and dryer felt and drawn under the roll. His head and chest were crushed and his right arm terribly burned .. .. .. the body stopped the machinery. He was scarcely eighteen years old.[7]

> September 7, 1909 — employed as a second hand.. .. .. was caught by the driving shaft beneath calendar number 22 at the papermills at three o'clock this morning and was instantly killed. He was assisting in replacing a belt and it is supposed that his hand was caught between the slack of the belt and the revolving shaft, and in an instant his body was thrown around the pulley and hurled with tremendous force against the iron rods which controlled the brakes. The machinery was stopped as soon as possible and when the body was removed it was found that one leg and one arm had been torn from the body .. .. .. he was but nineteen years of age.[8]

"None of you will prosper unless the Company prospers."

One can understand the workers' need to organize on the basis of just trying to survive. But if that wasn't enough, the paternalism of the employer was there to add fuel to the fire as illustrated in a letter the International Paper Company sent to employees dated May 2, 1904:

> None of you will prosper unless the Company prospers. The success of the Company will bring permanent employment to each of you, opportunity of advancement and an increased income.
>
> You are fortunate in that most of you have had steady employment for the past year and that you have all enjoyed the highest rates of pay which are received by any one in the communities in which you live.
>
> In closing, let us hope that you will all exercise moderation and will give the Company the loyal support which is necessary in order to bring the coming year to a successful issue.
>
> Yours truly,
>
> THE, INTERNATIONAL PAPER COMPANY

This letter followed a letter sent to employees the year before setting forth the company's views on the "labor situation as it existed then." Of course neither letter mentioned the contract the company signed with the union in 1902 and how that contract helped workers to "prosper." The attitude of the corporation, expressed in these letters, brings to mind the warning made by the Knights of Labor in its 1878 Declaration of Principles:*

> The alarming developments and aggressiveness of great capitalists and corporations, unless checked, will inevitably lead to the pauperization and hopeless degradation of the toiling masses.
>
> It is imperative, if we desire to enjoy the full blessings of life, that a check be placed upon unjust accumulation, and the power for evil of aggregated wealth.

Labor's Day

The managers of the company wrote these letters in response to a strong class consciousness that existed in the community and the mill. Workers were not just fighting among themselves, as illustrated by the problems between the Pulp & Sulphite Union and the Paper Makers Union. They were also agitating for higher wages, better working conditions and democracy both in the mill and in the community. This class action and consciousness was demonstrated by the political muscle labor exercised in the passage of a law that required the state to keep track of and report industrial accidents and in the union's 1902 contract which required the company to get union permission before running the mill on Sundays. It is also captured in a *Lewiston Evening Journal* September 8, 1903 article on the first labor day parade in Livermore Falls:**

> Monday will go down in history as the first time Labor Day was ever observed at Livermore Falls by representatives of the various labor unions;.. .. .. It was estimated that from 1,500 to 2,000 were in line.
>
> At an early hour the people began to fairly pour in from surrounding towns, and long before the arrival of the big special

* By 1886 the Knights of Labor had a national membership of 1 million with 17,503 in Maine where they met in 118 local Assemblies (Report of the Maine Commissioner of Industrial and Labor Statistics 1887, 157). The Knights were an inclusive union that represented the working class. A group of trade unions advocating a narrow agenda split off from the Knights that year and formed the American Federation of Labor.

** The very fact that Labor Day was a legal holiday was an indication of labor's strength. In 1891, due to pressure from organized labor, the Maine Legislature passed a law declaring that "the first Monday in September of each year, being the day celebrated and known as labor's holiday, is hereby made a legal holiday. . ."

> from Rumford Falls and from Lewiston, on the P. & R.F. Railway, the streets were just crowded with people.. .. .. The special was made up of four passenger coaches and ten flat cars on which seats had been arranged.. .. ..a reliable source said fifteen hundred people were aboard.

Chisholm Responds

When thousands of people showed up in Livermore Falls to celebrate Labor's Day, is it any wonder that the International Paper Company responded by telling its employees that they would be successful not by the sweat of the worker's own brow and organization, but only as the sweat of the worker's brow supported the company.

Oliver Twist Gives His Master a Shock

Paper Makers' Journal. February 1907.

Hugh Chisholm, founder and president of the International Paper Company, wrote an article on the history and status of the paper industry which was published in the 1906 report of the Maine Commissioner of Industrial and Labor Statistics. At the end of the article Chisholm warns:

> Another disturbing element is uncertainty as to how far the workmen in our mills will follow unwise leaders. The possible injury to our industry in this way is almost incalculable and no little burden rests on our newspapers to see to it that all workmen are fully advised of the results of intemperate acts. I am not counseling rabid, anti-union utterances, but I am pointing out the wisdom of putting one side of the question before the working man as clearly and forcibly as the other side is put at union meetings. If he can be made to think instead of to follow blindly, the danger will be over.[9]

Since Chisholm was the president, from 1898 to 1907,[10] of the largest paper corporation in the world, and a major economic force in Maine,* his directive to the newspapers and workers was taken seriously especially when you take into consideration Mr. Seitz of *The New York World's* testimony to Congress; "It is understood that the publisher of The World must print nothing hostile to the interests of the International Paper Company."

Counterpoint

But there was a counterpoint to Chisholm. Take, for example, the 225 women paper bag makers employed by him in Rumford, Maine who,

> Late in August (1906) went out on strike to enforce a demand for an increase in wages of 25 cents per day. So large a contingent going out necessarily obliged the mill to shut down. The strike lasted about three weeks, with a loss in wages, to employees, of about $30,000. The girls finally went back on the offer of the company, which was a slight increase over former wages.[11]

Newspaper publishers were also responding to the power of Chisholm and the International Paper Company. A number of newspapers including *The New York Times* had a spokesperson testify in front of the Judiciary Committee of the U.S. House of Representatives in 1904 that: "Moreover, we have reached the point where the [paper] trust is beginning to dominate opinion and to edit newspapers."[12] The "trust" he was referring to was the International Paper Company.

---

* International Paper Company in 1906 had many mills in the Northeast and in the Maine towns of Livermore Falls, Gardiner, Jay, Chisholm, Riley, Orono, Rumford, and Solon.

Endnotes

1. "Labor Day Report." *Lewiston Evening Journal.* September 8, 1903.
2. Burke, Letter to Harvard University.
3. *Papermakers' Journal.* Vol. 1. No. 3. 16.
4. *Annual Report 1895* 221.
5 Burns 83.
6. *Annual Report 1909* 225.
7. *Annual Report 1909* 225.
8. *Annual Report 1887* 244.
9. *Annual Report 1906* 169.
10. *The History Factory* 112.
11. *Annual Report 1906* 82.
12. *New York Times.* April 8, 1904. p. 8.

LABOR IN THE EARLY TWENTIETH CENTURY: A PRELUDE TO THE STRIKE OF 1908

4

CONDITIONS

In 1908 there was no National Labor Relations Board, no minimum wage, no Social Security and no unemployment insurance; indoor toilets, electricity and running water were scarce. Maine had a few labor laws. They were described as the laws that protected women, children and the working classes. These laws defined the age at which children could start working and limited the hours that women could work in industry.[1]

In the early 1900s workers had political power which they used to pass pro-labor laws, but most of the laws were declared unconstitutional by state and federal courts. Even child labor laws were successfully challenged, for a while, on the basis that they infringed upon the rights of "equal parties" to form contracts. There was no law in Maine preventing the permanent replacement of strikers, or lockouts by employers. There was an 1891 Labor Conspiracy Law which stated:

> Whoever by threats, intimidation or force, alone or in combination with others, prevents any person from entering into or continuing in the employment of any person, firm or corporation, shall be punished by imprisonment not more than two years, or a fine not exceeding five hundred dollars.

All the while corporate managers intimidated workers with private police, lockouts, lawyers, blacklists and scabs, as their judges routinely issued injunctions against union picket lines, strikes and boycotts.

BUREAU OF LABOR STATISTICS

The corporations of 1908, as now, relied on their power and wealth to tell their side of any story. But unlike today, the state, after careful investigation issued a report which analyzed the underlying causes of the strike and made the report part of the public record.

We can thank the Knights of Labor for these state reports because the Knights knew that if unions wanted to make broad societal changes they needed institutions that assembled and analyzed data on behalf of workers. They saw the establishment of state bureaus of labor statistics as vitally important. After much political agitation by the Knights in Maine, a law was passed in 1887 which created the Bureau of Industrial and Labor Statistics. This law required the state to keep track of and issue reports on the condition of labor in Maine. Thanks to these reports, we have a record of what happened in the strikes of 1908 and 1910.

Bargaining Units

Since there was no National Labor Relations Act, the government didn't have a role in defining union bargaining units. Rather, the bargaining units were defined by the power relationship between unions and employers with the help of the AFL.

Now we think of a union primarily as a collective bargaining agent which derives its power from government recognition. Recognition is determined by the outcome of an election in which workers are asked to vote for or against union representation. These election campaigns are run like political campaigns, except that in this case the company is one candidate and the union the other. The outcome of the election determines the union's status. If the union wins, the company and the workers have to recognize it. If the union loses, not only does the company not recognize it, in most cases neither do the workers.

In 1908 workers looked to their own organization and solidarity to move their agenda forward, not to government recognition through a certification election.

Safety Net

Most union contracts did not include concepts like seniority and benefits like health insurance, that we have today.* Skilled workers moved from company to company on a regular basis, and semi-skilled workers quit the mills quite often to work in the woods. The mills shut down in slow periods and there was no unemployment insurance or welfare to fall back on. When workers were injured or sick, they didn't work and they didn't get paid.

* There is no mention of seniority in the first contract between IP Co. and the Federal Labor Union in 1902 and the 1937 contract with the Pulp & Sulphite states, "If employees are to be promoted or laid-off, the management will take into consideration, first, ability, second, length of service."

Family, neighbors, church and the union were the safety net. And if they failed, the town provided the poor farm and the city the poor house.

POLITICS

Politics was different in the early 1900s because workers saw the world differently. Their organizations offered clear alternatives to a world in which the rich got richer and the poor got poorer. Things were not set in stone as they seem to be today, when labor tacitly accepts the premise that the rich will lead the country and the corporations will dominate all institutions in the society. Back at the dawn of the twentieth century many labor leaders were active in Socialist politics. John Burke, President of the Pulp & Sulphite Union from 1917 to 1965, said in 1908:

> .. .. .. every stand of the working class against the criminal greed and avarice of the capitalist class is a step forward in the inspiring march of the laboring masses towards their final complete emancipation. Brothers at Mechanicsville, why not strike at the ballot box next November? An industrial strike is crude; a political strike is scientific. Is there any sense in going on strike against a class that rides on our backs, that lives in our blood, and then go to the ballot box and vote that class into office? .. .. .. The working class must have members of their own class in every office in the nation.. .. ...[2]

Burke went on to lead the New Hampshire Socialist Party as its candidate for Governor in 1914; three years later, he was elected National President of the Pulp & Sulphite Union.

Workers did have some political victories during this period in Maine. In 1908, working people amended the constitution of Maine to allow citizens to bypass the corporate-led legislature and Governor to create laws through the citizen-initiated referendum process. As a result of labor agitation, a law was passed in 1909 requiring the State to keep and publish records of industrial accidents. These were important gains many don't consider today because, although we may have televisions and personal computers, most workers don't see the issues as clearly as workers did at the turn of the century. It was not because they were economically poorer than we are today that they saw the issues more clearly. It was because they had their own vision of the future from which to view the present. And they had institutions, like political parties, schools, newspapers (500 Labor newspapers in 1902) and unions, that were constantly talking about how to make that vision a reality.

CAPITAL OR LABOR?

But the biggest difference was the role workers played in the raging debate between labor and capital to determine who would lead the society. When people referred to capital they meant a concentration of both money and power in the hands of a few, like the property Hugh Chisholm owned in the paper industry and the power he wielded in society. By labor, people meant the masses of people out of whose labor capital is created. President Lincoln framed the issue in his first State of the Union Address, 1861:

> Labor is prior to and independent of, capital. Capital is only the fruit of labor, and could never have existed if labor had not first existed. Labor is the superior of capital, and deserves much the higher consideration. Capital has its rights, which are worthy of protection as any other rights. Nor is it denied that there is, and will probably be, a relationship between labor and capital producing mutual benefits. *The error is in assuming that the whole labor of the community exists within that relationship.* (Italics added)

The big question being debated in the early years of the century was in whose interest would the society be run; who would set the parameters of the political debate; who would define education; would the culture be consumer or cooperative? Capital or labor?

THE QUESTION OF THE HOUR

Paper Makers' Journal. April 1907.

ENDNOTES

1. Public Laws of the State of Maine 1887. Chapter 139. 121-125. "An Act to Regulate the Hours of Labor of Women and Children."

2. Voelker 2.

A GREAT LESSON: THE STRIKE OF 1908

5

Between 1906 and 1909 a rivalry took place between the two unions. Each raided the other for members and broke one another's strikes. Such rivalry, which extended into Maine, was one of the major ingredients of a long strike in the industry in 1908 . . . The papermakers were defeated in the strike which left them in a weakened condition. [1]

The Great Northern Lockout of 1907

As a backdrop to the 1908 strike against International Paper Company, a quick look at the 1907 lockout by the Great Northern Paper Company helps set up the problems that existed at the time between the Paper Makers Union and the Pulp & Sulphite Union.

Prior to the expiration of their contract in July, the Paper Makers at the mills of the Great Northern Paper Company in Maine requested an end to Sunday work and a wage increase of 25 cents a day. The corporation agreed to end Sunday work but would not agree to a wage increase. Several meetings were held over several months to negotiate wage levels for a new contract, but no agreement was reached. Then in November, the corporation locked-out the Paper Makers Union. The Pulp & Sulphite Union not only continued to work, but helped the corporation find people to replace the locked-out Paper Makers.

After ten days the managers of Great Northern ended the lockout when the Paper Makers agreed to work under the terms of the old contract for one year. But a major jurisdictional problem still existed between the company and the two unions. Today, the National Labor Relations Board decides what the scope of a bargaining unit will be. Back in 1907, the company, the AFL and both national unions agreed the AFL would determine the lines of jurisdictional demarcation between the two unions at Great Northern Paper Company.* This was done, but then the local Paper Makers Union rejected the agreement worked out by the AFL. The company responded by declaring its contract with the Paper Makers Union void. It refused to recognize the

* Back then the AFL was a signatory to the contract along with the company and the union.

Paper Makers Union, but continued to recognize the Pulp & Sulphite Union and acknowledged its jurisdiction over some jobs previously held by Paper Makers.

The Strike of 1908 *

The economy was not good, the market for paper was weak and the warehouses of the paper corporations were full when International Paper Company served notice on the Paper Makers Union that the corporation planned to cut wages by seven and three-quarters percent on August 1, 1908, the day after the contract expired. The Union responded that "The cost of living has increased eleven percent in the last two years.. .. .. we respectfully decline the proposition."[2]

The Paper Makers Union was informed that the International Paper Company Board of Directors would take up the union's response at the board's next meeting. After the Board meeting nothing was conveyed to the Paper Makers Union. So the Union set a strike date of August 2, 1908 to affect all the company's mills where they previously had contracts, including the Maine mills in Rumford, Livermore Falls, and Orono.

The timing of the strike was bad for the Paper Makers because with few orders and a big inventory, it was to the corporation's advantage to just shut down the mills when the strike began. Another problem was that the Paper Makers Union had acted alone, striking without the support of the Pulp & Sulphite Union. When the corporation shut down the mills in response to the Paper Makers Union strike, it also put the Pulp & Sulphite Union members out of work, and in so doing widened the gap between the two groups of workers. Finally, given the depressed market and large inventory, the strike saved the corporation from taking the heat for inflicting large layoffs. International Paper Company could now claim that it wanted to keep people working, albeit at a reduced rate, but the Paper Makers Union forced the corporation to shut the mills down.

As the strike went into its second month, the national President of the Paper Makers Union, J. T. Carey, signed an agreement with International Paper Company on September 24 which included a five percent reduction in wages. Under the agreement, foremen and night watchmen would not be recognized as members of the labor organization; the company agreed not to discriminate against strikers, and the union agreed to go back to work on September 28. But Carey had made the same mistake he made in the Great

---

* This strike was against both International Paper Company and Great Northern Paper Company, but for the sake of simplicity, only the International Paper Company strike of 1908 will be followed.

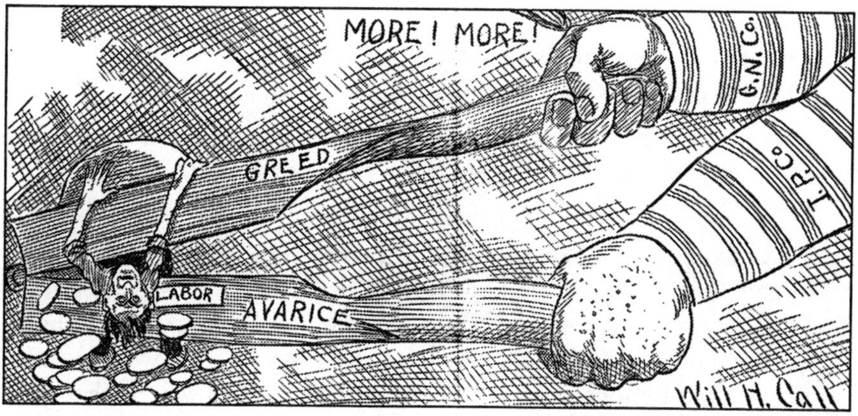

Paper Makers' Journal. October 1908.

Northern lockout the year before: he signed an agreement without the support of the local unions.

Today, local unions have no legal recourse to go against a decision of the national union. If they do not live up to a contract signed by the national union, the local union is often placed in receivership and new local leaders are appointed by the national leadership. And if the national union does not act to enforce a contract, the corporation may sue the union for any losses due to disruption of production.

But in 1908 things were different. The local unions responded to Carey by refusing to go back to work and informed him he had no right to make any settlement without first putting the company's proposal to a referendum vote of the membership. Maine workers not only insisted on democracy through citizen-initiated referendums in their State Legislature; they also insisted on democracy in their unions. In fact the Paper Makers Union went even further when at their 1902 convention they adopted the initiative and referendum system of legislation, and election of officers for themselves.[3]

Demands

The Paper Makers Union met in convention on September 29, 1908, in Albany, New York, and submitted a proposal to the International Paper Company that included the following:

- That all employees of the company be in good standing with their respective unions according to the jurisdiction of the American Federation of Labor.
- That all members of the Paper Makers Union be reinstated in their former positions immediately upon the mills' resumption of operations.

- That wages be not less than those in effect as of June 15, 1908.
- That hours of work remain the same.
- That boss machine tenders and foremen would remain in the union.
- That further curtailment of production would be brought about by stopping paper plants entirely for a fraction of the week instead of running part of the machines continually.[4]

It is clear from these demands that after two months on strike the workers still wanted to fight for some basic principles that their National President had abandoned. They wanted foremen in the union. They wanted to spread the pain and protect all the members by substituting a shorter work week for layoffs. But most importantly, *workers* wanted the right to vote on the agreements that *they* had to live under.

International Paper Company Responds

In 1908, the International Paper Company responded to the offer from the Paper Makers Union conference the same way Great Northern Paper Company had the year before. They opened up the mills and decreed that because the national union could not convince the local unions to abide by the agreement, International Paper Company would not recognize the Paper Makers Union. The corporation advertised for strike breakers and, when they arrived, used them to scab the Paper Makers' jobs.

Defeat

Meanwhile, the Pulp & Sulphite Union members returned to their own jobs and the mills started to run again. On November 7, 1908, the local Paper Makers Unions, acting this time on instructions from the national union, called off the strike and accepted the conditions previously offered by the corporation. The Maine Bureau of Industrial and Labor Statistics reported:

> Many of them found their old positions occupied by unfamiliar faces and they will have to seek employment elsewhere or in some other industry. This is one of the regrettable consequences of a strike that terminates in this way. The Paper Makers after a determined fight of fourteen weeks for what they considered principles upon which depends the life and existence of their organization return to work defeated but not discouraged or disorganized.[5]

Contrary to the report of the Commissioner, the members of the Paper Makers Union were discouraged and disorganized. The members of the Pulp & Sulphite Union were also discouraged and disorganized even though they had not participated in the strike. Correction: they did participate. They fought on the wrong side. Here is how the situation was reported to the membership in the *Paper Makers' Journal.*[6]

> During upwards of three years there has been almost constant clash or contest as to jurisdiction of those two unions, which had become so bitter that during the strike of the Paper Makers' Union in 1908, the Pulp, Sulphite and Paper Mill Workers' Union actually assisted the employers in filling the places of strikers, with the result that the strike was a failure. There-upon the employing corporation executed a new trade agreement with the Pulp Makers' Union, which practically gave the first named organization jurisdiction over the entire paper making operation.

After the strike of 1908 membership in both unions dropped to almost nothing. But the news wasn't *all* bad.

Trade Secrets

Remember that in 1908 workers did not hold jobs by seniority and they did not look to the corporation for lifetime employment. When they left a job they took all their skills with them. Today when a worker leaves a job many of the skills stay in a computer or in a machine. Not in 1908. Without seniority systems and the expectation of lifetime employment, and with the ability to take their trade secrets with them, Paper Makers often moved from mill to mill. A former president of a Machine Tenders Union in the early part of the twentieth century said: "I sit by the fireside now and think of the old days when the machine tenders roamed the country. No new machine could be started without them. They were kings."[7]

A RAGGED COURTSHIP
(A Pro-Convention Idyl)

Paper Makers' Journal. June 1907.

Closed Shop

Aiding union workers in their ability to move from mill to mill was the concept of the "closed shop." Many of the contracts unions negotiated during this period required the company to hire only people who were already members of the union. This concept is called the closed shop. The closed shop clause in the 1907 Pulp & Sulphite contract with the Great Northern Paper Company stated:

> In consideration of the Great Northern Paper Company employing .. .. .. only members of the above named organization, for operating at its mills at Millinocket, East Millinocket and Madison, the union named herein, and the officers for said Union, **will furnish** the Great Northern Paper Company, for its mills, when requested, with all the necessary skilled and unskilled workmen.. .. (bold added)

The closed shop was a powerful tool in building and maintaining union strength. It helped the union develop industry-wide apprenticeship, wage and working conditions standards and find jobs for workers who were laid off, locked out or fired for labor activity.

The closed shop has been illegal since the passage of the Taft-Hartley Act of 1947.* Today the best a union can do is negotiate a "union shop," which means that the employer can hire anyone, but after a time stipulated in the contract (30 or 60 days is the norm), the employee must join the union. During that probationary period the union has no legal grounds to represent the worker. Thus the corporate managers decide who is hired, thereby denying the union a powerful tool they once had in the workplace and the community.

Unit Size

Today, workers are primarily restricted to organizing units authorized by the National Labor Relations Board.** The Board usually determines unit size based on a request of the union. Due to lack of resources unions usually request that a bargaining unit represent only one facility or mill. This works fine if the company only owns and operates one facility. But in a large company the union is immediately placed at a great disadvantage in bargaining if it has to represent many relatively small units whose contracts expire at different times. This puts the small unit in a hole when it has to square off against the huge giant of a corporation by itself. Workers in Mobile, Jay, Lockhaven and Depere faced this situation in 1987 when they took on the

* For all practical purposes the Building and Construction Trade Unions are closed shops where they have contracts.

** There are exceptions, like farm workers, who are not covered by the NLRA, public sector workers who are covered by state laws and other federal laws, and interstate transportation workers who come under other Federal acts.

International Paper Company which still had contracts in place with local unions at 22 other primary mills and many secondary mills as well.

Union Standards / Industry Wide

The unions fought to set common industrial labor standards that workers took from mill to mill. Union standards stopped corporations from whip-sawing workers between different facilities and different companies. They put a stop to a corporate-sponsored race to see who would work for the lowest wages and under the worst conditions.

The closed shop and union standards also gave the union a place in workers' lives far more important than present-day unions. Several years ago, a very pro-union great-grandmother in Livermore Falls told me a story about her father, a union mason, who moved to Livermore Falls around 1910. He had the reputation of being the best mason around and the manager of the paper mill wanted the mason to build a large fireplace in the manager's new home. The manager was in a big rush and told the mason he wanted him to work 12-hour days and through the weekend. The mason said he was union, and that union masons work eight-hour days and he would rather not work weekends but would if paid double-time.

Standards today are something unions try to negotiate *for* workers. Years ago standards were created and upheld not just by the official union apparatus but by union members. This activity of members upholding standards helped give union workers an identity which they took with them everywhere.

Unity Out of Disunity

The union, as we understand a union today, was destroyed in the strike of 1908. However, the union people had learned a great lesson: *divided we fall.* But the demise of the unions was only temporary because the working class culture that existed then was like a perennial plant that sleeps and rests all winter. It emerged in the spring of 1910 to create a vibrant union organization and move the labor agenda ahead for workers and their communities in the paper mill towns for the next decade.

Endnotes

1. Scontras 49.
2. *Annual Report 1908* 422.
3. Burns 85.
4. *Annual Report 1908* 425.

5. *Annual Report 1908* 429.
6. *Papermakers' Journal.* July 1910 6.
7. Gross 185.

THE GREAT LESSON APPLIED: THE STRIKE OF 1910

6

Even the strike breakers who had been hired in 1908 now had reason to complain of the company's failure to carry out work contracts which had enticed them originally into 'scabbing.' The men suffered all kinds of intimidation with no collective means of self-defense.[1]

After the strike of 1908 Thomas J. Lyons, the Maine Commissioner of Industrial and Labor Statistics, reported: "The Paper Makers as an organization have no prestige in the mills of the International and Great Northern Paper Companies, and so far as they are concerned, the mills are open shops."* [2]

What did the people think about one union that broke another union's strike? That is what happened in 1908: Pulp Workers broke the Paper Makers' strike. So how did two unions that almost destroyed each other in 1908 move on and join together as one to win a strike two years later? A strike whose outcome set the industrial standard for working conditions that all paper mill workers would strive to maintain for the next decade.

The 1908 Paper Makers' strike was called off in November of that year and a contract was signed with a 5 percent wage reduction and a no-discrimination clause against hiring union workers. However, the companies paid scant attention to the contract: "Snubbing the Union ... IP and Great Northern, immediately violated the 'no discrimination clause' of the agreement by frequent use of blacklists and 'yellow dog contracts,' and by starting Sunday operation of the mills." ** [3]

The actions of the companies forced the unions to look carefully at their past activities and the following year, in February 1909, 30 delegates of the Paper Makers Union met in Boston to talk over their problems. The Pulp & Sulphite Union followed with a convention in May, and on June 10, 1909

---

* An open shop is a non-union shop. An open shop contractor or company is one that will go to any extreme to keep workers from organizing a union; an open shop has "a union-free environment."

** "Yellow dog contracts" are contracts that individual workers are required to sign with employers stating the worker will never join a union. Union organizers could be jailed for conspiring to breach a yellow dog contract (Hitchman v. Mitchell 245 U.S. 229). Yellow dog contracts are illegal today under the Norris -LaGuardia and National Labor Relations Acts.

the two unions signed a truce. The AFL, at the request of the Paper Makers Union, recognized the Pulp & Sulphite Union shortly thereafter on July 31, 1909.

However, the differences between the two unions still ran deep and hard. The Paper Makers were highly skilled and were often shop floor bosses. The Paper Makers had often pushed their own agenda at the expense of other workers, causing them to form a separate union. The Pulp & Sulphite workers were no angels either. They crossed picket lines and scabbed the jobs of the Paper Makers in the strikes of 1907 and 1908. The good news was that both unions learned from their mistakes and together called a strike at the International Paper Company mill in Corinth, New York, on March 6, 1910. The issue was the discharge of a union brother for attending a conference: the mill manager had given him time off, but while the brother was attending the meeting, management at corporate headquarters ordered him fired. That was the spark that started the strike of 1910.

Organizing the Strike

Workers without a formal organization in many of the International Paper Company mills came together in secret. The 1948 Documentary Papers of Maurice Jones explains:

> How the strike was organized 1910. We had 48 men paying dues to headquarters. We had a secret committee of three of these who did not know each other until after the strike. A retired carpenter and myself went over a list of names of 20 men — all were true blue. The carpenter selected three names from the list. They communicated via the carpenter — slipping notes under his door at night. They had an underground with Glens Falls, Fort Edward and Ticonderoga and Albany. The Company had what was known as '$1,500 a month men' — traitors who advocated reorganization to find out who were active. What happened to them is still a mystery.. .. .. The underground with people connected with the IP Company gave us the conviction that we could strike and win. This was our foundation on which to build.[4]

Outside View — the Cause

After interviewing strikers, meeting with the executive committees of the unions in Corinth, Fort Edwards and Ticonderoga, and talking to some of the International Paper Company mill superintendents, the Commissioner of the New York Department of Labor, John A. Fitch, issued a preliminary report on the causes of the strike on April 21, 1910: "The most important causes appear to be the issue of unionism — the right of the men to an organization and the recognition of such an organization." The Maine Commissioner of Industrial and Labor Statistics blamed the company for the 1910 strike:

> The strike in the paper mills in 1908 resulted in a defeat for the employees and they returned to work practically disorganized in many of the mills. Notwithstanding this, they were assured by the officers of the International Paper Company that the company had nothing against the Union and that if the men would perfect their organization and put themselves in a position to be recognized, the company would willingly do business with them. The men claimed that the promise was not kept and that a spy system was maintained in the mills under which it was impossible for them to reorganize. Men were discharged without any cause being given and when they applied for reinstatement they were required to sign an agreement to the effect that they would not join any labor organization while in the employ of the International Paper Company. These extractions on the part of the company created a condition that could not continue for a great while and the discharge of a man at Corinth was made the occasion for a determined protest.[5]

Prelude

We can imagine hundreds of clandestine meetings and discussions among workers where their mistakes, leading up to the disaster of 1908, were hashed over and over.* Finally they agreed to put their grievances aside and work together on local and national levels. Then, they waited for the opportunity to come out in the open and take on the International Paper Company. But while they waited the union people had to quietly defend themselves as best they could from the company's spy system: "Spies were placed in every mill; their job was to spot the active union men and report to their leader. The company set up what was known then as the blue-card system. When a man was marked for dismissal, he found a blue card attached to his time card."[6]

The organization in many of the mills was still weak or non-existent in March of 1910, but a new strategy based on lessons learned the hard way over the previous decade soon put the workers on the offensive and the company

* The reason it is necessary to imagine what was going on is because the Pulp & Sulphite Union papers from 1909 and 1910 are missing. The problem for us is not that the facts are missing, but the culture in which the events took place is also gone. If we had documents written by the people who planned and participated in these events, it would give us some insight into that very culture. It is that culture, one in which working class people believed they could shape their society, that we need to reclaim.

on the defensive. Up to this point the organization the workers had been building in the International Paper Company system had of necessity primarily been secret. Where before organizing had been slow, methodical and behind the scenes, it now became fast, exciting and public. With the walkout at the Corinth mill, the struggle to "perfect the organization" took a new and dramatic turn.*

Let us now follow the organization that spread the strike, perfected the organization, and built in the process two strong national unions in the pulp and paper industry. When reading these accounts, think of how the lessons of 1910 could be applied to union strike and organizing strategies today. Look, also, at how two weak national unions used a strike to build two strong national unions.

The Strike Begins

Within a week of its humble beginnings in Corinth the paperworkers' strike of 1910 had spread to International Paper Company mills in Glens Falls and Fort Edward, New York. Some of the machines in both communities were being operated non-union and, "Legal moves, rather than violence, characterized the opening of the second week of the strike."[7] The company brought a court action asking for $100,000 due to losses sustained during the strike and in New York State a judge imposed an injunction preventing union members from picketing or interfering with strikebreakers. Meanwhile, President Carey of the Paper Makers Union announced in Watertown, New York, that the strike would spread to the Maine mills.

The National Guard

With the corporation and its supporters in government fearing the strike might spread to Maine, Company C, 2nd Regiment Infantry of the National Guard of the State of Maine, was called out to make a show of force by holding a drill in Livermore Falls on March 7. [8] **

* The term "perfect the organization" was used by union, government and company sources to define the building of organization.

** The Great Railroad Strike of 1877, which began as a local strike but spread like wildfire across 14 states in two weeks, was finally put down by Federal troops. By the time the dust settled more than 100 people had died and the railroad owners were convinced of the necessity, from their point of view, of the formation of State National Guard Units to put down labor revolts.

It is interesting to note that this National Guard unit was organized in Livermore Falls on May 15, 1908, just prior to the strike that began in August of that year.

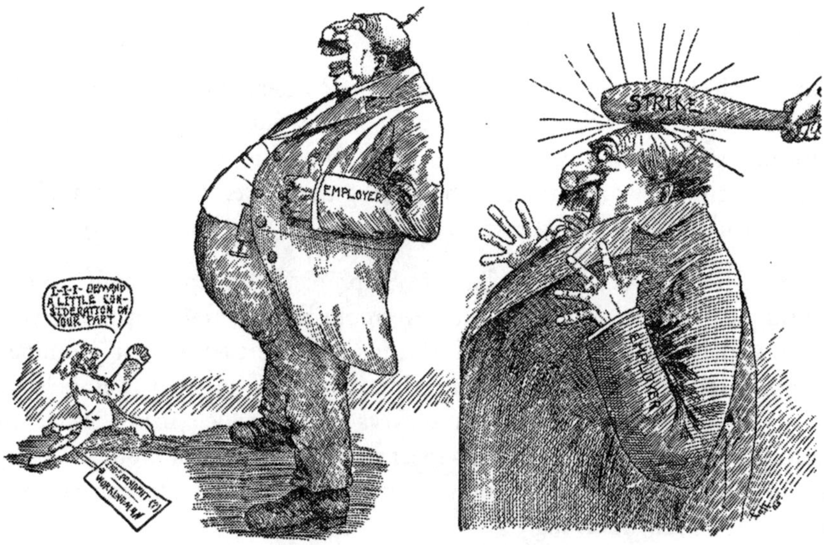

Individual Efforts versus Organized Accomplishments

Paper Makers' Journal. December 1905.

The Company's Position

International Paper Company managers disputed the strikers' analysis of conditions in the mills. They released a statement, reprinted in the *Lewiston Sun* on March 14:

> We believe misunderstanding of our motives and our attitude toward you, probably in some measure brought about through misrepresentation, is largely responsible for the present strike at several of our mills. In the first place we emphatically state that we severally and collectively have only the kindliest feelings toward you and that we are really solicitous for your welfare.
>
> We would remind you that our stockholders are almost as numerous as our employees and that a large proportion of them are quite as dependent for their living expenses and comfort upon the income from their investment in this company as you and your family are upon your wages. A large portion of our stockholders are women, many of them are widows. The strike of 1908 inflicted a severe loss upon these people.
>
> You cannot force the issue, you can only retard it by untimely demands. We invite not only those who are now in our employ, but former employees, to tell us as individuals what fault they have with our treatment of them.[9]

Four More Go Down

On March 20, President Carey of the Paper Makers Union announced that four more International Paper Company mills would not be opening in the morning and the mill in Livermore Falls, employing 600 men running 10 paper machines, would probably be joining the strike. It was estimated that 2,500 workers were then on strike against the company.

John Malin, President of the Pulp & Sulphite Union, spoke to more than 1,000 people at the Music Hall in Livermore Falls.[10] Present at the meeting were workers, their families, their friends and the town's most prominent businessmen. After the public meeting, a private meeting was held for International Paper Company employees during which many joined the union. That evening another mass meeting was held: a strike vote was taken, a general strike was approved and workers began leaving the mill shortly after midnight Sunday morning.

By the next day, most of the International Paper Company employees in Chisholm, Riley and Livermore Falls, Maine were on strike. Meanwhile, in Portland, Maine an agent for the company was hiring scabs for the struck mills in New York. "The Company stopped hiring strike breakers a week ago after shipping some 300 men, but has resumed" hiring.[11]

While President Carey of the Paper Makers Union was warmly greeted in Livermore Falls, things were not so good up the river in Rumford. President Malin of the Pulp & Sulphite Union and organizer George Schneider of the Paper Makers were run out of town by a citizens committee and told not to return: "If Malin and Schneider return here tomorrow, harsher measures are likely to be used against them."[12]

The Pattern

That was the pattern the national unions used to organize the strike and keep it going. The leadership and activists from nearby communities came to town, organized a mass public meeting, urged workers to join the union and urged the community to support the union. Once organized, the new union would take a strike vote. As we will see later in Rumford, workers did not always vote to strike.

Public Sympathy? / Newspaper Bias

Back in Livermore Falls on March 23rd, the *Lewiston Sun* reported: "The strike has been in progress three days and those who anticipated immediate adjustment of the trouble have given up all hope for an early settlement."

The paper went on to state that two years earlier the citizens in town were in sympathy with the strikers, "but that is not true today." The paper claimed the strike was caused by national union leaders and the local workers only struck in support of the New York workers. A company spokesperson said Sunday work was abolished when their employees asked for the mills to be closed on Sunday, so that could not be an issue.

Interestingly, in 1987 the Governor of Maine, John McKernan, portrayed the strike against International Paper Company as a power struggle between Big Business and Big Labor. He was just as wrong in 1987 as the *Lewiston Sun* was in 1910. In most cases, people who speak for the corporate class find it hard to believe (or at least admit) that the cause of strikes is usually a power struggle between working class needs and corporate class greed. In 1910 and 1987, it was local workers who voted to strike.

It is hard to acknowledge intelligent human capacity in people that you plan to force into submission, because in doing so you call into question your own ethical involvement in the process. We harvest vegetables without remorse because we do not grant vegetables the attributes of intelligence or feelings. It is easier to kill a chicken than it is a cow because everyone knows chickens have very tiny brains and birds are just lower on the evolutionary scale then mammals like cows. It is easier to suppress the just aspirations of workers if you can convince yourself that they are struggling for their rights not because they know what they want, but because someone else is telling them what to do. If you can convince yourself that workers are too stupid to exercise power on their own, then you can convince yourself that it is okay to defeat them by any means. And this becomes a principle.

"It's a Great Principle"

The principle is the right of an employer to prevent, by any means, the formation of unions by workers. John D. Rockefeller evoked this principle when he testified in 1914 at congressional hearings investigating a strike in which state militia, acting on behalf of a Rockefeller company in Ludlow, Colorado, fired on and killed 14 people, including 11 children. Rockefeller responded to questions from committee members:

> Question: And you would do that if it cost all your property and kills all your employees?
>
> Rockefeller: It is a great principle.
>
> Question: And you would do it rather than recognize the right of men to collectively bargain? Is that what I understand?

> Rockefeller: No, Sir. Rather than allow outside people to come in and interfere with employees who are thoroughly satisfied with their labor conditions.[13]

Rockefeller justified the killing of 14 people because he was upholding the principle of his right to prevent workers from organizing. One has to wonder how Rockefeller would have felt if workers had murdered his children and then used the principle of their right to organize as the justification.

The Strike Goes On

Returning to the Paperworkers' strike, President John Malin of the Pulp & Sulphite Union moved on from Rumford to Franklin, New Hampshire, where the City Council adopted a resolution asking him not to stay because a riot might start if he addressed the mill workers.

> Malin said that the vote of the council would not hasten his departure from the city. He said that he merely wanted to lay the New York situation before the union members and that was entirely within his rights. It is understood that when he leaves this city he will go to Berlin, N.H., where the Company has a number of mills.[14]

Malin spoke to a meeting of workers in Franklin, but no strike vote was taken. Meanwhile, on the same day organizer George Schneider returned to Rumford. "The labor people here are indignant at what has been done and if the citizens continue this form of procedure there is likely to be serious trouble here."[15] It was also reported that "many of the labor men here are beginning to rally around the organizer .. .. .. and that today 40 persons have joined the union." Schneider stayed in Rumford for awhile to help workers perfect their organization, then "it will be up to the union, my work will have ended and the question of whether to strike would have to be decided by a vote of the union."[16] He added: "Any further action on the part of the so-called citizens to deport me from town will be immediately rebuked by the labor men and you will see a strike here within 24 hours."

Momentum Builds and Workers Set the Record Straight

On March 24, workers at the International Paper Company mills in Solon, Maine and Franklin, New Hampshire, struck. The Pulp Workers' press committee in Livermore Falls issued a statement: "We are greatly pleased to say that the company started three paper machines this morning, sorry they did not make any paper to ship." They noted that during the 1908 strike many rumors spread by the company were believed by the strikers, but this time: "This committee have charge of news, therefore all news published are facts.

By so doing we enable all to know just what is going on in this and other mills. The old schemes won't work this time, we are all wise."[17]

On the twenty-second day of the strike, workers at the mill in Ticonderoga, New York walked out and now 10 of International Paper Company's 32 mills were idle. Workers at the Milton, New Hampshire mill joined on March 28. The company's Board of Directors met at the end of March and passed a resolution directing company officers to continue their present attitude until all the mills are in full operation. That same day workers in Careyville, New York responded to the directors by voting to strike and shut down two of the town's three mills the next day. Despite the directions of the Board of Directors, the company's managers made a strategic move in Franklin, New Hampshire on March 30. They boarded up three of the six mills and shipped the scabs to other mills International Paper Company was trying to run.

The union was now organizing faster than the corporation. On April 3, 1910, workers in Berlin, New Hampshire held a mass meeting and voted to strike. John Malin, Pulp & Sulphite President who had been run out of Rumford two weeks before, spoke at that meeting. While Malin was in Berlin, President Jeremiah Carey of the Paper Makers Union was in Livermore Falls speaking at a mass meeting. Carey then went up to Rumford and addressed a meeting there.

The energy created by the strike affected paper mill workers at other companies. "Organizer Doughten arrived here (Lisbon Falls, Maine) from Rumford today and is engaged in perfecting the organization of the employees of the plants controlled by the Bowdoin Paper Manufacturing Co."[18]

What was really going on? *

It is known from the *Paper Makers' Journal* that there was a lot of communication between workers from the different mill towns. The constant movement of workers from mill to mill over the years created human bonds of trust and independent lines of communication that went deeper then the formal lines of the union structure. In analyzing the period, anthropologist August Carbonella concludes:

---

* The internal union papers of the Pulp & Sulphite Union for the years 1909 and 1910 are missing from the labor archives at Cornell and the University of Wisconsin. These two schools house the early papers of both the Paper Makers Union and Pulp & Sulphite Union. Neither Archive can explain why the papers for 1909-10 are missing. The librarian at the UPIU headquarters in Nashville, Tennessee, has no explanation for the missing papers either. The Pulp & Sulphite Union was the larger and more aggressive of the two unions. The Paper Makers' papers make little reference to how the organizing was actually done. Most of the information on the day-to-day activities of the 1910 strike I have taken from the *Lewiston Sun* newspaper which had correspondents in the Maine paper mill towns of Rumford and Livermore Falls. They also reprinted wire stories emanating from the mill towns in other states. The *Lewiston Sun* was very anti-union in its reporting, so it is necessary to read between the lines to glean information on actual organizing activity.

> Another regional labor practice — union control of hiring — greatly enhanced this labor culture. This control facilitated the circulation of workers in and around the region, and gave them a measure of informal control over their working lives. This migratory practice significantly built up a network of intimate relations throughout the region and even beyond. This expanded spatial perspective created an intimate sense of class belonging and consciousness, a perspective both sustained by and sustaining of the regional labor culture.[19]

We know from the newspaper accounts of mass meetings that the leaders were constantly moving from town to town, educating and organizing workers during the strike, or in their words "perfecting the organization." But what were the actual workers on strike doing? The following article printed on April 6, 1910 in the *Lewiston Sun* gives some insight into the anti-union bias of the paper and a glimpse into the activity of Livermore Falls strikers.

THE GRIP OF GREED AND HAND OF GENEROSITY

Paper Workers' Journal. June 1906.

**LABOR LEADERS LEAVE
RUMFORD CITIZENS BELIEVE ORGANIZERS
HAVE GIVEN UP HERE.**

> For the first time since March 21, Rumford is tonight free from labor organizers. This morning President Carey left on the train.. .. ..
>
> The party of labor agitators who came here from Livermore Falls on Tuesday morning by automobile, started for home shortly after breakfast hour today. Thereby leaving Rumford free from labor leaders and the citizens in a happy frame of mind. Conditions remain quiet here and the general opinion here expressed is that labor organizers have given up the fight here.

On April 10 organizer George Schneider returned to Rumford and took up residence in the Hotel Rumford. Although the workers in Rumford never did strike in 1910, the national unions continued to help them perfect their organization. (After the strike the company did recognize the unions in Rumford.)

By the middle of April, the press reported that the 7,000 striking employees had lost $60,000 a week in wages and the company had lost half a million dollars.[20] The company claimed it would be back in full operation by the middle of May and promised:

> Company stores to provide employees with supplies practically at cost and company boarding houses will be established at points where citizens are not disposed to accord fair and impartial treatment to men who come back to work for the company.[21]

Corporate Tactics

Corporations often try to split working class communities with gifts. On April 14 Hugh Chisholm, former President of the International Paper Company, met with community, business and professional men in Rumford and offered to help them set up a non-denominational YMCA. He showed them drawings of a three-story structure that would be built across from the hotel and would house a dining hall, lecture room and gymnasium. The building still stands in Rumford today.

Other tactics traditionally used by corporations against unions during strikes that were employed in this one were suits for damages against the unions (i.e., for lost production); injunctions against picketing; a show of force by police and the National Guard; use of scabs; closings of mills; having the press play up division and violence where none existed; having the press ignore positive union activity; and paying off union officials to sell out the membership. All these tactics and more, including outright murder, were in common use at the time, but the working class fought back.

On May Day 1910, workers all over New England were on strike: 4,200 trolley men in western Massachusetts, 2,500 building trades people in Rhode Island and Massachusetts, and many others joined the paper mill workers on picket lines of their own that spring.

VICTORY

On May 4 , the article below appeared in the North Jay News column of the *Lewiston Sun*:

> **NORTH JAY NEWS**
>
> **ORDINATION OF MR. NEIVEEN NEXT THURSDAY—STRIKE SETTLED SATISFACTORILY**
>
> North Jay, May 4. — This much feared strike was settled satisfactorily Saturday night. The new bill is signed by the company for five years, at a rise in wages with 10 per cent the first year and 20 per cent the next four years.
>
> Woodbine Temple Pythian Sisters had for entertainment last evening, the role call followed by a short farce entitled "How the Story Grew,".. .. ..
>
> *Lewiston Sun*. May 5, 1910.

Thus, the strike was settled. However, the article in the *Lewiston Sun* left out a few details. The unions won a grievance procedure, the elimination of Sunday work, the eight-hour day and most importantly they had put their own house in order. The *Lewiston Sun* didn't report any celebration of the great victory, but one can imagine there was some partying going on when the unions transformed the International Paper Company mill towns into union towns in May 1910.

The Lessons of 1908–1910

The paper mill workers made a number of mistakes in the strike of 1908. They struck when business was slow and warehouses were full. After the strike began, the union couldn't gain momentum because the work force was divided. The next mistake was the signing of an agreement with the company by the national union leadership that the membership didn't accept. Finally, when the company reopened the mills, the managers were able to take advantage of the divided work force. In short, the papermakers learned the meaning to the second part of the old union credo, "United we stand, divided we fall." This class division not only broke the strike, but also destroyed the credibility of both union organizations in the eyes of many workers.

United We Stand

In 1910, workers experienced the other side of that old union credo. The leadership and membership of two national unions combined forces and worked actively to build local solidarity and power bases. They learned from their mistakes, put their differences aside and used the strike to build their unions. By the time the strike of 1910 was over, the unions that barely existed in the International Paper Company mills in 1909 had prestige, power and influence in the mills and the towns that surrounded them. This victory lasted a decade.

One important lesson of 1910 is that at the beginning of the twentieth century, unions had the ability to learn from their mistakes. They were self-critical and so could admit their mistakes, analyze them and move on. Today, self-critical labor institutions are suppressed because they are seen as threats by those who hold the $150,000-a-year jobs at the top of the labor bureaucracy. Staughton Lynd, writing for *IMPACT* in 1998, states: "Our purpose in not speaking out is to safeguard the labor movement from external attack, but the result of our doing so is to destroy the labor movement from within."[22] It is impossible for us to plan for the future if we do not have any institutional mechanisms to analyze the past and the present; and on top of that, we are afraid or made to feel uncomfortable when we state our honest opinions.

UNITY

It is also my contention that the major problem facing the working class is not the external forces that are arrayed against it, like the military might of armies or the concentration of economic power held by people of the corporate class. The major problem is the inability to create unity in the ranks. Whenever workers fight among themselves, they lose. But when even a semblance of unity is obtained, the opposition often falls. That is why words like "solidarity" and "unity" and sayings like "An injury to one is an injury to all" mean so much. Those words did not fall out of the sky to the union people in the paper mill towns during the first decade of the twentieth century. They learned about disunity, and the pestilence it sows, in the devastating defeat of 1908, but they were able to put their differences behind them and unite to create a victory in 1910 that lasted a decade. Could they ever have done that if they weren't constantly critiquing their own actions?

In August of 1987, president Wayne Glenn of the Paperworkers Union was scheduled to participate in a UPIU Region 1 staff meeting in Maine. En route to the staff meeting, he spoke at the weekly mass meeting held in Jay.* These meetings were open to the public and attended by at least 1,000 people, week in and week out, for 16 months. Several weeks before, Glenn had turned down an invitation to speak at a rally and march in Jay which was attended by over 10,000 people. He claimed he had conflicting duties. After attending the Jay mass meeting he did speak once to the workers locked out by the International Paper Company in Mobile, Alabama and to meetings of workers on strike against the company in Depere, Wisconsin and Lockhaven, Pennsylvania. There were UPIU locals in 22 other International Paper Company primary mills and many more at the secondary mills in the United States. During the course of the strike, to my knowledge, he never spoke at *any* of their meetings.

However, workers from Jay did go and speak to workers at every primary and most of the secondary International Paper Company locations. Jack Getman, in his book *The Betrayal of Local 14*, goes into great detail explaining how the national union in 1987/88 not only gave little support to Local 14's attempt to spread the strike but in fact did much to undercut that effort. Getman supports his claim by raising the following questions:

- Why did the union acquiesce so easily in the break-up of the multiples?**

* The time of the mass meeting was changed to accommodate Glenn's schedule. This meant that the thousands of people who attended the meeting also had to change their schedules. Many had to leave work early to be able to attend.

** The reference here is to the two units or "multiples" the company bargained with beginning in 1937. One multiple was all the primary mills in IP's Northern Division and the other was the Southern Kraft

- Why was so little done to convince the Pine Bluff locals to join the pool?
- Why did the International not notify .. .. .. strike leaders when the reps for Hamilton and Corinth let their contracts expire without giving notice?
- Why was the cover story, that IP had requested talks on condition that the corporate campaign be stopped, used to explain the Louisville negotiations?* Was it a way of getting rid of Ray Rogers? (See Chapter 13, Labor's New Sit-down)
- Why did the international not seek a boycott of International Paper?
- Why did it not pursue the environmental issue more aggressively?
- Why, after the strike, did Wayne Glenn (international union president) give assurances that the corporate campaign would continue, shortly before it was turned off?[23]

Had Glenn and his staff gone from mill town to mill town organizing and holding public meetings, the way it was done in 1910, who knows what might have happened? What if they had stopped the breakup of the multiple bargaining unit to begin with? What if they had a program to bring the Pine Bluff locals into the pool? What if they had served notice in Corinth and Hamilton, called for a boycott of International Paper Company early in the fight, and pursued the environmental issues, the corporate campaign and coordinated bargaining more aggressively?

The leadership and the membership of 1910 knew they were fighting the whole corporation, and to beat that corporation they had to unite a critical mass of the people who labored for that corporation. Even if the workers at one mill, such as Rumford, didn't strike, the union kept the pressure on the company and never gave up on the workers.

Tragically, as we shall see, the lessons of 1910 were not passed on to the people who inherited the paper mill workers' unions after 1937.

Summary

The strike of 1910 was the greatest victory ever for paper mill workers. People who grow up and live in paper mill towns should know the story of the 1910 strike, not because they should be trying to duplicate that strike, but because it is a proud moment in their history and it can teach them much about themselves. Finally, if workers do not learn from their losses and pass on and celebrate their victories, their struggles become a hopeless whirlpool

---

Division. By the mid-1980s, with the exception of pension councils, the company was bargaining only with locals one location at a time.

* Getman argues that the national negotiations between International Paper Company and the UPIU were precipitated by Wayne Glenn as a face saving measure for the union to end the strike, not as a victory for the strikers. In return for the company participating in the negotiations the union had to end the Corporate Campaign and get rid of Rogers.

of defeat. Their destiny, like that of the pet mouse always running fast but getting nowhere on the inner track of a Ferris wheel, will be to repeat the mistakes of the past over and over and over again.

ENDNOTES

1. *A Group of Men* 26.
2. *Annual Report 1908* 429.
3. Burns 149.
4. *A Group of Men* 26.
5. Fitch.
6. Gross 188.
7. "Sues Union." *Lewiston Sun.* March 14, 1910.
8. "At Livermore Falls—Paper Mills Quiet—Inspection of Company C." *Lewiston Sun.* March 18, 1910.
9. "Paper Co To Old (sic) Employes." *Lewiston Sun.* March 14, 1910.
10. "Livermore Falls Strike—Big Mass Meeting of Paper Makers Yesterday Afternoon." *Lewiston Sun.* March 21, 1910.
11. "Hiring Men In Maine." *Lewiston Sun.* March 21, 1910.
12. "Leaders Talk." *Lewiston Sun.* March 22, 1910.
13. *Who Built America.* Vol. II. 197.
14. "Malin At Franklin N.H." *Lewiston Sun.* March 23, 1910.
15. "Organizer Schneider Is Guarded By Policemen." *Lewiston Sun.* March 23, 1910.
16. "Strikebreakers Stoned at Livermore Falls. Schneider Still at Work in Rumford." *Lewiston Sun.* March 24, 1910.
17. "A Union Statement." *Lewiston Sun.* March 24, 1910.
18. "Organizer Doughten at Lisbon." *Lewiston Sun.* April 6, 1910.
19. Carbonella 90.
20. "To Sue Rumford Officials." *Lewiston Sun.* April 14, 1910.
21. "Paper Co's Statement." *Lewiston Sun.* April 12, 1910.
22. Lynd.
23. Getman 222.

Union Membership by Year in Unions Nationally and in the Paper Industry 1909-1920			
Year	**Paper Makers**	**Pulp & Sulphite**	**All Unions**
1909	1,000	1,000	2,000,000
1910	1,600	700	2,200,000
1911	2,400	2,800	2,300,000
1916	5,200	4,400	2,800,000
1917	6,400	6,500	3,100,000
1918	6,000	8,000	3,600,000
1919	5,500	8,400	4,300,000
1920	7,400	9,500	4,800,000

Note: Data for different unions are not strictly comparable—provided for trends only.

7

CONSOLIDATING POWER: THE SLIPPERY SLOPE–1910 to 1921

August Carbonella, in his paper "Historical Memory, Class Formation, and Power," sums up the impact of union activity in the Jay area during the ten years that followed the strike of 1910:

> *This victory generated a kind of organizational euphoria throughout the region. The emerging strength of the labor movement enabled unions to funnel funds, material aid and organizers into the area. Labor reports for Jay show a steady increase in the number of workers represented in all sectors of the local economy up to 1920. The AFL council hall increasingly served as the political, cultural and intellectual center of the community during this period of activity. Weekly meetings were generally packed. One member of the pulp/sulphite union remembers people 'standing on the stairs and even overflowing out into the streets trying to listen to what was going on.' Consumption and leisure were also deeply politicized in this rush to organization. The labor council organized a consumer co-op, a baseball league, weekly dances as well as a lecture series. The sports competitions, in particular, served as a means to ritualize tensions between skilled and unskilled workers — between*

> *Irish-Americans and French-Canadians — creating a fragile unity that would be broken in the strike of 1921. But for a short time at least, the union movement created a political culture in which the social divisions of the working-class were subsumed by a more expansive social vision.*

John Burke, President of the Pulp & Sulphite Union said in 1916: "When the 1910 strike came along, that strike saved this organization. If it had not been for that strike, and if we had not won that strike six years ago, there would not be any trade union movement in the paper making industry at the present time."[1] The union people took the momentum of their victory and used it to improve the content of their lives and the lives of others. They channeled aggressive behavior to the ballfield and creative behavior to the stage plays they put on. They set up a Central Labor Council where problems between unions were resolved and transformed into a unity that confronted the company when management threatened union standards. The negative energy that in the past had been directed at the Pulp & Sulphite Union by the Paper Makers Union, and vice versa, now was combined into a positive energy used to promote the well being of the working class community.

COMMUNICATIONS

Paper Makers Union Local #11 reported in the March 1911 edition of the *Paper Makers' Journal*: "Well, we made the people of Livermore Falls stand up. We had to, no room for them to sit down, 500 in a hall that will hold 360" for a play put on by the Central Labor Union. The *Paper Makers' Journal*, a 16-page report of local and national union news, grew in increments after the strike into a monthly journal of 40 to 60 pages. It is a sad commentary on the present state of union communications to say that there isn't even one union monthly publication that comes close to what the Paper Makers Union had in that period, and there are no labor newspapers to match the 500 that existed in 1902. In fact, not one major newspaper in this country has a labor page, and to add insult to injury, labor news is often printed in the business section, written by reporters who cover business news.

UNION PRIVILEGE CARD OR COOPERATIVE STORE?

Lane Kirkland, President of the AFL-CIO in 1995, was criticized for heading a labor movement that wasn't moving. Kirkland responded by saying, "This movement has not been standing still, it has not been ignoring the problems and needs."[2] In defense of his record, he said he fought to get minorities and women named to the executive council of the AFL-CIO and pushed for a program called the "union privilege card," whereby union members can get special credit cards.

If Kirkland's crowning economic achievement was to enable union members to get a special interest rate on a credit card, it raises an embarrassing question: has this resulted in one person joining a union? And if someone did, would they quit the union if some corporation offered them a better deal on another credit card? Suppose Kirkland had instead spearheaded a drive to set up a national union bank in which union members could help each other by pooling their money and making sure mortgages were available to working class people. If this bank, with its pool of money, was used as a source of power to really create good union jobs (as has been done in other countries), then Kirkland would have had a record to stand on and would have won respect for himself and for all of organized labor.

The Central Labor Union in Livermore Falls in 1916 did not make credit cards available so union people could get discounts on goods made by prison labor in China or child labor in India. It set up a co-operative food store and a powerful union label committee to make sure stores in town sold union-made goods.

Move It or Lose It

In 1903, 2,000 people marched in the Livermore Falls Labor Day parade. In 1994 there wasn't a Labor Day parade in the entire State of Maine. In 1987 and 1988, when strikers from Jay were crisscrossing the state, educating, organizing and mobilizing union people, thousands showed up for Labor Day celebrations. *Union activity is contagious. Union inactivity is deadly.* The term "use it or lose it" is aptly applied to unions. Working people need to build active unions that are engaged in promoting the welfare of the working class both inside and outside the workplace. If not, then the institution of the union will either die or be allowed to survive as a means of controlling workers rather then promoting the interests of the working class.

Local Changes – National Changes

The Two Wilsons

During the second decade of the twentieth century labor's power on a national level was reflected in the election of the two Wilsons, William and Woodrow. Woodrow Wilson was President of the United States from 1913 to 1921. William Wilson was born in Scotland and worked in the Pennsylvania coal mines as a child. He helped organize the United Mine Workers of America and served in the US Congress from 1907 to 1913. William Wilson helped draft the bill that created the US Department of Labor and then served under President Woodrow Wilson as the first Secretary of Labor from 1913 to 1921.

Industrial Feudalism

Congress in 1913, responding to the growth of organized labor and what some members of Congress considered its radical and violent activities, set up a Commission on Industrial Relations to examine what some thought was the problem, namely, radical unions. Chairman Frank Walsh held hearings to expose the underlying cause of labor unrest: the concentration of wealth in the hands of autocratic management, a system he referred to as "industrial feudalism." The Commission issued its final report in 1915 and called for an inheritance tax to finance education, social services, and public works. It called for laws to protect the rights of workers to join unions and bargain collectively and it called for equal pay for men and women. Walsh shows up several years later as co-chair of the War Labor Board where he was successful in implementing some of these recommendations during the war years.

Labor's Legislative Agenda – on the Books

During President Wilson's first term some important parts of labor's legislative agenda were enacted into law. The US Constitution was amended in 1913 to replace the appointing of US Senators with election by popular vote. This was followed in Wilson's second term with the Constitutional amendment granting women the right to vote.

A law passed in 1914, the Clayton Act, exempted labor unions from antitrust action under the Sherman Act of 1890. The Sherman Act was passed after much agitation against the corporate trusts that had been set up to run and control the newly emerging giant industrial concerns. The political activity against the corporate trusts had been led by the labor, progressive and populist movements of the day. However, the Supreme Court ruled that unions could at times be considered labor trusts under the Sherman Act. The AFL declared that amending the Sherman Act to exclude union activity was its top priority and considered the Clayton Act "labor's Magna Charta." However, the courts did not see things the way the AFL did and, "In ten of the thirteen decisions reported between 1916 and 1920 .. .. .. the Act was held not to impede the granting of an injunction" against labor.[3] In other words the Clayton Act did little to stop judges from granting injunctions against labor utilization of the 13th and 1st Amendments.

In 1915 the Seamen's Act was passed which eliminated the debt peonage system for American sailors. Until its passage a sailor could be forced to continue working for a corporation as long as he owed it money. Think of it — if you worked and lived on a ship at sea you had no choice but to shop at the "company store" and pay whatever price they asked. Then when you went into debt paying the exorbitant prices and interest rates the company charged, you were restrained from leaving the ship until your bill was paid.

A federal child labor law was passed in 1916, later declared unconstitutional. The first federal law was passed regulating the hours of work for private sector workers; the Adamson Act granted railroad workers the eight-hour day. All of this happened not because Wilson was a nice guy, but because workers in communities all over the country (like Jay and Livermore Falls) were in motion, "stirring the pot."

Debs and Wilson

The Socialist Party candidate for president in 1912 was Eugene Victor Debs, a nationally known labor leader who helped build the American Railway Union, had been involved in the great uprising of 1877, led the great railroad strike of 1894 and help found the Industrial Workers of the World (I.W.W.) in 1905. Debs received 900,000 votes for president in 1912, while Theodore Roosevelt got 4,100,000 running on the Progressive ticket; Taft, the Republican incumbent, got 3,484,000 and Wilson won with 6,293,000. Wilson's record in his first term earned him the support of progressives and many unionists deserted the Socialist Party to vote for Wilson when Debs didn't run in 1916.

Wilson's 1916 victory was due in part to labor support and his campaign promise to keep the United States out of the war being waged in Europe, a promise he did not keep. Just 29 days after starting his second term, Wilson asked Congress to declare war, which it did. Debs urged people to oppose the war, and for this he was indicted under the Espionage Act in 1917, found guilty and sentenced to 10 years in jail. In 1920 he ran for president from jail and received 919,799 votes. The Democrat James M. Cox lost; the Republican Warren G. Harding won and the progressive legislation of the previous decade was rolled back the same way the Republicans in the 1980s and Democrats in the 1990s have rolled back much of the progressive legislation passed under the New Deal of Franklin D. Roosevelt.

War

Debs, the I.W.W. and the Socialist Party opposed the war because war generally is not good for the working class. War is the most extreme form of competition. World War I was being fought to see who would rule Europe. It was not being fought to increase the minimum wage, end child labor, or bring about universal free education, health care as a right or equal pay for equal work. While war is the ultimate competition in which workers from one country fight workers from another to the death, it does have a few benefits:

Labor and the War

> .. .. .. A great labor 'turn-over' has taken place. December 1, 1916, there were engaged in manufacturing in the state of Maine 99,665 persons. December 1, 1918 there were employed 113,410. Thus there were 13,745 more people employed in manufacturing at the close of the war than at the beginning. When it is considered that about 25,000 men of Maine were mustered into the service of the United States during this time, the question will arise where the additional number engaged in manufacturing came from. .. .. .. 29,988 people engaged in manufacturing in the state were employed directly on war orders at the time the armistice was signed.[4]

War eliminates competition for jobs among workers inside the countries at war and in some of the countries allied with the warring countries. War also has a tendency to put a lid on the population explosion and working class women get a shot at jobs they were previously told they were incapable of doing.

In Jay there is living memory of how this worked during World War II and it may have been similar during World War I.* During World War II, skilled male workers too old to be drafted left the paper mills for higher paying jobs 60 miles away at the Bath Iron Works shipyard, where Navy ships were being built. Because most of the young men were going into the military, many of the skilled and unskilled paper mill jobs were taken by women. When the war ended the women were laid off as the men came back from the military and Bath Iron Works. Not surprisingly, at that time, many articles appeared in national publications "elevating" the status of women as housewives on a pedestal in an attempt to "disappear" the image of "Rosie the Riveter" that had been promoted during the war to encourage women to take industrial jobs.

Also during World War I, the great wave of European immigration came to a standstill. This also contributed to a more favorable labor market for many US workers.

Corporations and War

War is good for corporations in the long run and bad for the working class. For one thing, corporations do not die in wars. In fact, corporations are not flesh and blood, but are really legal fictions. So corporations cannot die a natural death as soldiers and civilians do in war. Instead, their owners usually make out like bandits. Why?

* There is significant data on women in the workplace around the country during WWI which supports my thesis of what likely occurred in the Jay paper mills (*Who Built America.* Vol. II. 220, 244).

Corporations are institutions that consolidate wealth and power, and that is what war is usually about. The US entered the war in 1917. In came wage, price and production controls and out of it came the suppression of union and political activity which heralded the beginning of the end for the radical unions that were part of the American labor movement.

WHIPSAW REVERSED

The First World War proceeded to move European workers from the workplace to the war fronts while workers in the United States picked up the slack for them as more and more jobs were created here. More jobs meant less competition for those jobs, and so corporations were competing for workers instead of workers competing for jobs. When jobs are scarce, the corporations whipsaw workers by threatening to move or close plants if lower wages are not accepted. When corporations have to compete for workers, the whipsaw process is reversed and workers can demand and get higher wages. But to make sure workers' wages did not get too high and to prevent strikes that affected war production during World War I, the Federal Government set up the War Labor Board.

THE WAR LABOR BOARD

During the war government support of workers in war-related industries was in sync with government strategy to promote political democracy overseas by promoting industrial democracy at home. As a result labor organizations that promoted the war increased their membership and power.

In order to prevent labor-management disputes from curtailing war-related production, President Wilson created The War Labor Board co-chaired by former President Taft and former Commission on Industrial Relations Chair, Frank Walsh. This Board conducted workplace elections to decide not which unions — but which workers — would form a workplace council that would represent the workforce. Disagreement between these councils and management were arbitrated by the Board. If a corporation failed to go along with a Board decision, President Wilson used his war time powers to nationalize the corporation. On the other hand, if workers failed to abide by a Board ruling the President sent the troops.

Although the War Labor Board didn't promote mainstream unions per se, the reality was that union candidates ran for the worker representative positions. These union people invariably won because they represented the best organized part of the work force. Hence, as union power increased, so did union membership.

The democracy advocated by the United States government during the war was hollow. When Congress declared war in 1917, it effectively abolished free speech by enacting the Espionage Act, which made it a crime to say anything that would "discourage" enlistment in the armed forces. The American Federation of Labor supported the war and its unions flourished.

However, members of the Industrial Workers of the World (the I.W.W. or "Wobblies") expressed opposition to the war. They were harassed and jailed, and as a result, their union was seriously weakened. One of the jailed Wobblies was Ralph Chaplin, who wrote the lyrics to Labor's anthem "Solidarity Forever." While in jail, Chaplin wrote a book of poems called *Bars and Shadows.* In the introduction, economist Scott Nearing explained why the Wobblies were singled out for persecution:

> Long before the war, the I.W.W. had made itself known and feared for its conduct of strikes; its free speech fights, and its ability to put the sore spots of American industrial life on the front pages of the daily press...It was in the domain of industry that the I.W.W. was functioning, and it was among the business interests that the determination had been reached to rid the country of the organization at all costs.[5]

The best-known of the thousands jailed for violating the Espionage Act was labor leader Eugene V. Debs, who helped organize the American Railway Union and had been jailed earlier for leading the Pullman strike of 1894. Debs was sentenced to 10 years in federal prison under the Espionage Act for making anti-war speeches.[6]

The war years of 1917 to 1918 were both good and bad for unions. The working class was empowered by the unions' call for industrial democracy here and by the government's call to "make the world safe for democracy" over there. And President Wilson would not allow the corporate managers to wage war against workers who belonged to unions that went along with the war. However, once the war ended, all bets were off and the corporate managers declared war on the unions. This time the government enlisted on management's side and with the I.W.W. out of the way, the AFL unions took it on the chin.

Paper Mill Strikes of 1918 and 1919

On average, a million people went on strike every year from 1916 through 1920.* These strikes happened even though the Federal Government had set

* The average number of people who went on strike a year from 1996 through 2000 was 300,000. The population in 1915 was 100 million and in 2000 it was 276 million.

up boards to determine wages and prices. The Maine Commissioner of Labor and Industry reported on one of those strikes:

> Employees at the International Paper Company at Rumford and Livermore Falls went on strike late in July (1918) to enforce a decision by the Federal Wage Board as to the interpretation of a clause of its award affecting the back pay due employees. A conference was held and a sub-committee appointed composed of members of the Wage Board, which reported that while the International Paper Company was not legally bound to pay this bonus it was morally bound to keep faith with its employees.[7]

On April 26, 1918 a case regarding wages and unionization of some 19 companies producing newsprint was brought before the War Labor Board. On June 27 the Board issued its report requiring, among other things, a 10 cent an hour increase; wages to be adjusted for inflation every 6 months; 8-hour day for workers in the mills and 9 hours for those outside with both getting time and one-half for overtime; females to get same pay as males for doing the same work; four shut down holidays; and:

> the rights of the workers to organize in trade unions and to bargain collectively through chosen representatives is hereby recognized and affirmed. This right shall not be denied, abridged, or interfered with by employers or their representatives in any manner whatsoever.[8]

Of all the employers only the International Paper Company refused to implement the Board's award precipitating a strike which forced the closing of mills in Glens Falls, Fort Edward, Corinth and Ticonderoga, New York and Rumford, Maine on July 24, 1918. The following day workers in Livermore Falls and other locations joined the strike as a conference was held by the War Labor Board to hear both sides. On July 26 the Board issued an "Interpretation of Award" and the *New York Times* reported:

> . . .that the employees were entitled to an increase of 10 cents an hour and a shortening of the workday from nine to eight hours, but they were not entitled to a 10 per cent bonus granted by the company before the award became effective early in July.[9]

The 4,000 strikers were ordered back to work. In Livermore Falls, 750 workers voted to end the strike and the mill opened on July 29.

The controversy stemmed from an increase the company granted on October 21, 1917 which included a 10 percent bonus. In a letter the International Paper Company informed employees: "This Company has decided to still further share the burden of its employees in meeting the cost of living, by again making a readjustment and increase in the wages now paid all hourly employees." The company went on to say that the increase will be a "serious burden" on them, but: "At the same time, it realizes that abnormal conditions, caused by the existing state of war, require unusual considerations of its employees .. .. .. and expects this action on its part will ensure the hearty co-operation of its employees."[10] In effect the Board ruled that the company had to grant a 10 percent increase but didn't have to honor its 10 percent bonus of October 21, 1917. The Board did suggest that the company was morally bound to honor its bonus.

However the controversy continued and the case was appealed. An umpire was chosen on December 6, 1918. He set a hearing date for January 1919 but then resigned. A new umpire was named who held hearings and issued a report approved by the Board on January 28 which reaffirmed the Board's previous order. But the story doesn't end here.

The Strike of 1919

> May 12, 1919 — Refusing an offer made by the International Paper Co. of approximately ten percent increase in wages in a new working agreement which was to go into effect today, 5,000 members of the International Brotherhood of Pulp, Sulphite and Paper Mill Workers employed in the company's mills thru-out the country struck.[11]

The War Labor Board informed the Union that it must abide by the Board's ruling which was effective until the end of the war. The union responded that the war ended on November 11, 1918. The Board took the position that the war wouldn't be over until the Treaty of Peace was ratified. The strike continued for two weeks and on May 26, "The men returned to work under orders of the union officials, pending a conference with the company over wage demands."[12]

The Treaty of Peace was signed on June 28, 1919 at Versailles, France and the union signed a one year contract with the International Paper Company 10 months later in New York on April 14, 1920 which included the following:[13]

- Union Shop
- Jurisdictional question to be resolved by the American Federation of Labor
- Mills will run six 24-hour days a week and shut down on Sundays
- 8-hour shifts with time and one-half for overtime
- The mills will shut down for July 4th — 32 hours; Labor Day — 24 hours; Christmas Day — 40 hours

- No strikes or lockouts
- Grievance procedure ends in binding arbitration

With the expiration of the contract in May 1921 war broke out between the unions and the International Paper Company. The company won and refused to recognize the defeated unions for the next 16 years.

SUMMARY

The years from 1910 to 1920 were the glory years for the paper mill workers and their communities. They survived a World War. They learned how to organize and exercise power. And they used that power to protect themselves from arrogant corporate managers and to promote a working class agenda of their own. But when World War I ended, government power that had been used to promote production and keep the capitalists off of labor's throat was reversed, and labor blood started to flow.

ENDNOTES

1. Voelker 82.
2. *Palm Beach Post.* February 21, 1995 3.
3. Forbath 158.
4. *Annual Report 1918* 10.
5. Ralph Chaplin 8.
6. Howard Zinn 16.
7. *Annual Report 1918* 17.
8. National War Labor Board. Docket No. 35.
9. "Paper Mill Strikers Ordered to Return." *New York Times.* July 28, 1918.
10. Company Notice of October 18, 1917.
11. "Strike of 5,000 I.P. Paper Mill Workers." *Lewiston Evening Journal.* May 12, 1919.
12. "Paper Mill Strikers Go Back To Work." *Lewiston Evening Journal.* April 26, 1919.
13. Contract information taken from Burke.

8

Union Membership by Year in Unions Nationally and in the Paper Industry 1921-1935			
Year	**Paper Makers**	**Pulp & Sulphite**	**All Unions**
1921	10,700	11,300	4,400,000
1922	8,300	6,800	3,700,000
1923	7,000	4,600	3,500,000
1935	9,300	15,700	3,700,000

Note: Data for different unions are not strictly comparable—provided for trends only.

THE STRIKE OF 1921: AFTERMATH AND LESSONS

In the early 1920s the manufacturers, aided by the post war depression, attempted to install the open shop. The struggle was centered in the mills and plants of the largest firm, the International Paper Company. The first pulp sulphite local was organized at an International Paper Company Mill. From that date until April 30, 1921, employees of IP composed the largest group of members in the unions. From 1911 to 1921, the pulp workers had 20 local unions in the mills of the IP, all covered by a blanket union shop agreement. By the spring of 1926, the pulp locals in the mills of International Paper had been destroyed, the union's survival made a matter of doubt, and from 1921 to 1937, all the mills of the International Paper Company were operated on an open-shop basis. *

In the aftermath of World War I, the lifting of wage, production and price controls was accompanied by the big corporations declaring an all-out war on the unions. One of the first battles took place in 1919 when 350,000 striking steel workers were defeated by the steel corporations with the help of thousands of private guards, state and local police, deputy sheriffs and judges issuing injunctions that suspended the civil liberties of the steel workers. The

* Taken from *The Paper Rebellion* (p.6), by Harry Graham, a book on the history of the split that took place in 1964 when 20,000 workers left the Paper Makers Union and the Pulp & Sulphite Union to form a more democratic union called the Association of Western Pulp and Paper Workers. In 1994 the leadership of the Association of Western Pulp and Paper Workers cut a deal with the leadership of the United Paperworkers International Union to affiliate. Ironically the membership of the Western Association rejected the deal and most locals went with the Carpenters Union.

defeat for workers in steel was followed by many others including those employed in the rail, coal, paper, and textile industries. Union membership dropped from 4.8 million in 1920 to 3.5 million in 1923. These massive union losses eventually led to significant changes in the way paperworker leaders viewed the potential role of their unions.

The American Plan

While the working class was solidifying and building institutions of its own from 1910 through 1920, International Paper Company and the capitalist class were getting ready to stomp out the growing labor movement and impose a ruling class social order. The corporations embraced what they called the American Plan and intended to substitute it for the economic, political and social institutions the unions had been developing. The American Plan, the corporation's idea of a private social welfare system for workers, called for the open shop, company-run welfare institutions, company-run social institutions and company-controlled local governments.

Concessions Rejected

In April 1921, the unions rejected the paper companies' offer of a 30 percent wage cut, a nine-hour day and the elimination of overtime pay. The contract's expiration date was May 1 in some mills and May 11 in others. The companies presented a united front. Their ideology, backed by their wealth and power, tested the limits of the union's organization. The companies' goal was to impose the American Plan. The workers' goal was to keep the power they had at work and in the communities where they lived.

Hitting the Bricks

Beginning on May 1 when the first contracts expired, and continuing through May 11, 25,000 paper mill workers shut down most of the paper industry in North America. The strikers held strong and before the end of the first month, the companies were talking about putting the strike issues in front of a panel of arbitrators. However, a split developed in the company ranks when the unions and all the companies, with the exception of International Paper Company, agreed to settle the dispute through arbitration. A deal was struck between the Carlisle Group (so named because it was led by Fred Carlisle, President of St. Regis Paper Company), and the unions. The Carlisle Group and the unions proceeded to assemble an arbitration panel. The panel issued its report, and in late June workers returned to all the mills except those of the International Paper Company.

The returning workers assessed themselves 10 percent of their wages to support those still on the front lines. However, that support only lasted six months when a spilt between the unions was orchestrated by the Carlisle Group with the help of the Paper Makers Union. This split played a major role in defeating the strikers at the International Paper Company mills.

UNION TACTICS

The tactics used by the unions in 1921 were similar to those used in 1910. The national leaders moved from town to town holding mass meetings. The difference this time was that the unions did not need to organize new locals. The locals had all been in place for a decade in cities and towns like Berlin, New Hampshire, where the city government was actively involved in the strike. At a mass meeting in Rumford, Maine, it was reported:

> The city clerk of Berlin, N.H., told of the situation in his city, as a result of the strike there. He said the people of Berlin were struggling to uphold the Union. The Mayor of Berlin, he said, had spent $175,000 of the city's money in improving the streets and sidewalks as a means of giving employment to the hundreds of men now idle on account of the strike. The people of Berlin, including city officials are squarely behind the strikers, he declared.[1]

The City of Berlin told the managers of International Paper Company that it would protect the company's property, but if the company tried to bring in scabs the city would do everything in its power to stop them. International Paper Company sued the city for $100,000 (a tactic it also employed in the 1910 and 1987 strikes).

CHRONOLOGY OF THE STRIKE

Throughout the summer of 1921 the strikers put up a fierce struggle in the paper mill towns. A brief chronology of the strike of 1921 follows:

- Workers in Palmer, New York hold mass public meetings every day of the week except Sunday. They do this for the entire five years of the strike.
- June 2 — With all the International Paper mills shut down, the company offers to return all strikers to work without prejudice if they accept a 21 percent reduction in wages.[2]
- June 21 — The wage conference between the unions and International Paper Company breaks up when the unions reject the company's wage reduction and the company rejects the union's offer of arbitration.[3]

- June 28 — Striking paper mill workers in the United States and Canada vote to accept the Carlisle Group's proposal of arbitration and return to work late in June. International Paper Company is not a party to the agreement.[4]
- July 2 — International Paper Company announces that it will open its mills on July 5.[5]
- July 22 — "During the entire time the men have been out of work it was hard to believe that any labor troubles were on in town, every one has worn a smile, but now the serious part is approaching and everyone is on a keen edge."[6]
    - Strikers attack a train bearing strikebreakers attempting to go into the International Paper Company mill in Corinth, New York[7]
    - The Livermore Falls Chamber of Commerce sent a letter on July 22 to International Paper Company urging it not to bring strikebreakers into Livermore Falls.[8] Similar letters were sent to the company from other Chambers and Town Councils. A typical reply said in part: "Former employees abandoned our mills and have refused high wages. Our mills will be started by willing, competent and peaceful men, whom we will deal with directly. .. .. .. your town will be held responsible legally.. .. .. I caution you to take notice of this fact. (Signed) International Paper Co."[9]
- July 26 — "Bellows Falls, Vt. — State troops have been mobilized for riot duty here in case the situation growing out of the strike of employees of the International Paper Co. should become critical.. .. .."[10]
- July 28 — "North Walpole, N.H. — Twenty patrolmen were sent to guard the International Paper Co's log yard.. .. .. following threats by the strikers they would dynamite the boom holding the logs in place"[11]
    - Labor leaders conferred with Maine Governor Baxter on the labor situation. "The Governor said he would do all his office would let him in aiding the men."[12]
- August 1 — The International Paper Company ordered that all persons living in Company houses in Chisholm (Jay) vacate them to make room for new workers who will operate the mills in Livermore Falls.[13]
- August 5 — 100 strikebreakers protected by Sheriffs arrived in Livermore Falls. An International Paper Company spokesman said the company would attempt to run the mill.[14]
- August 9 — "Livermore Falls — Half the hundred men brought here as strikebreakers today began operation of the sulphite mill in Chisholm. The other half are on guard duty."[15]
- August 15 — The International Paper Company announced it was closing the Rumford mill for six months and would start to move machinery to other mills.[16]
- August 25 — "Bangor — Sheriff Fernandez and a special posse have gone to Montague, in West Enfield, to protect the property of the International Paper Co. against rioters, said to be strikers. Several days ago two camps owned by the com-

pany were blown up with dynamite .. .. .. and large quantities of pulp have been thrown in the river."[17]

- August 31 — "New York — Another attempt will be made to settle the wage dispute in the paper mill industry when representatives of the International Paper Co. and the union meet."[18]

Labor Day 1921

Five months into the strike a Labor Day celebration was held in Livermore Falls. The headline in the *Livermore Falls Advertiser* on September 7, 1921 read, "BIG LABOR DAY PARADE MORE THAN A MILE LONG." The article described the parade as being a mile-and-a-half long, adding: "A most notable part of the parade was the large number of automobiles in line, **nearly every business man in town who owned an auto placed it in the procession.**" [emphasis added] The children of the town marched behind the band which followed the Chief Marshal who led the parade. With the children was a mule on whose sides were placards which read "All of our relatives worked in open shops." The article goes on to describe a full day of activities which included games of baseball between the "Irishmen and the Frenchmen" and a tug of war between the "Pulp & Sulphite and the Paper Makers." The day was rounded out with a presentation of the musical comedy "The Traveling Salesman," with the proceeds donated to the Central Labor Union.

A Line in the Sand

The late August negotiations went nowhere. On September 20, International Paper Company officials bought space in the local newspapers to proclaim ".. .. .. no negotiations of any character have taken place or will take place with the general or local officers representing the men on strike.. .. .. So far as the company is concerned all its services will be governed by the company's operating rules of May 24, 1921, together with the operating rules of the company dated July 5, 1921.. .. .."[19] International Paper had drawn a line in the sand. The president of the company clearly stated its position:

> The day for arbitration is passed. We can find plenty of men to operate our mills at the wage offered . . . Bolshevism does not control this country yet .....
>
> You will please understand very distinctly that this company proposes to run its own business. . . . and that it will not permit outside parties, directly or indirectly, to control its affairs.[20]

The company matched its rhetoric with its deeds as a paper worker recalled 50 years later:

"Well the thing to watch during that time was, ah, about every week there was a list of injunctions served by the court. You couldn't even walk the streets, you see. It was a violation of court order, and all you had to do was have somebody accuse you of violating the order and bang, you were in for thirty days."[21]

THE TIDE TURNS

The unions continued to put up a strong fight during the fall, with huge Labor Day celebrations, mass meetings, direct action and local political support. But some of the skilled workers started to drift back into the mills and the courts and state governments clearly took the side of the company. A Maine judge found 22 strikers guilty of conspiracy. They had been picketing the Rumford mill in violation of an injunction because, as the Judge said, the picket line interfered with the operation of the mill.[22] The national news was not encouraging either, as union leaders were put in jail or deported and strike after strike in other industries was lost.

The union people fought on and many stayed true, but the company started running machines and more and more workers crossed the line. Two front page articles in the *Livermore Falls Advertiser* on December 7, 1921, gave the mind-set of the parties at war. One article, based on union statements, was headlined:

STICK! IS THE WORD HANDED DOWN LINE

And a Large Body of Men Are Doing This Very Thing

The other article, based on company statements, began:

QUIETLY DRIFTING WITH THE EBB TIDE

Number Increasing Daily---Men Being Placed As Fast As Possible

The union claimed that Livermore Falls had been 100 percent union since 1910 and that after seven months 80 percent of the men are "sticking together with the tenacity of a French bulldog, and claim that before long victory will be theirs." The company article claimed that 304 men had returned to work and published their names — day by day.

A Scab Forever

The local paper continued to publish the names of the local people who crossed the line. Those who didn't cross were still on strike 66 years later. I was in a store in Livermore Falls one day in 1988, 62 years after the strike was officially called off, when an old man I was talking to mumbled something about another old man who had just entered the store. When the second old man left the store, I asked the first what the problem was. He responded, "That bastard is one of those who scabbed on us in 1921. He turned his back on us then, and I will never forgive him."

PARRY HAD A LITTLE RAT.

Paper Workers' Journal. December 1907.

THE COMPANY PLAYED THREE CARDS AND WON.

International Paper showed a profit in 1920 of $12 million, but it lost more than $7 million in 1921 and $5 million in 1922. Then in 1924, the company showed a profit of $2 million. The first two years of the strike occurred during a depressed period for the industry, but as the economy picked up the company couldn't take advantage of the market because of the strike. This gave the national leadership of the unions reason to think the strike could be won. However, three things happened that turned their optimism around. The first was that the Carlisle Group of companies wanted to reopen negotiations on wages; the second was a break in solidarity between the two unions; and the third was the operation of the Three Rivers Mill in Quebec.

The Carlisle Group

Under the arbitration agreement worked out with the Carlisle Group of companies, the wage settlement could be reopened at the request of either party. When the price of paper dropped, the companies re-convened the arbitration board to consider wages. The board was made up of three management members, three union members and one neutral member. On January 4, 1922, the Board voted, four to three, to accept a wage reduction of eight cents an hour for workers getting less than 54 cents an hour and no reduction for workers getting more than that. This meant that 90 percent of the Pulp & Sulphite Union members received a 20 percent cut in wages, while the Paper Makers Union members were untouched.[23]

In response to the settlement, John Burke, President of the Pulp & Sulphite Union, said, "Do not blame the rank and file of the Paper Makers in the Carlisle Group — place the blame where it belongs. You know where it belongs.. .. .. I believe they would have voted right on the proposition if they had been advised right."[24] The Paper Makers Union had once again broken faith with the Pulp & Sulphite Union and Burke pointed the finger.* Burke insinuated that the leaders of the Paper Makers should have taken the decision of the arbitration board to the members and ask that the members reject the decision even though the ruling of the arbitration board was not subject to approval by the membership. Burke believed the leadership of the Paper

---

* In this book for simplicity's sake I only refer to the two major unions in the industry. Please note that there were many other unions in the paper industry besides the Paper Makers and the Pulp & Sulphite. For example, in 1912 the following unions also had contracts with the International Paper Company: Stationary Firemen and Oilers, Steam Engineers, Plumbers and Steam Fitters, Carpenters, Electrical Workers, and Machinists. In the strike of 1921 many of the members of these unions crossed the picket lines with the Paper Makers in the second year of the strike.

Makers Union cut a deal with management. A deal which isolated the Pulp & Sulphite Union members in the Carlisle Group and stranded all the union members on strike against the International Paper Company. The wages of the highly skilled Paper Makers were not reduced, but the Pulp & Sulphite Union members took a huge hit. Alone.

Unity *Broken* — Pulp & Sulphite *Stranded*

This break in solidarity between the Paper Makers Union and the Pulp & Sulphite Union in the Carlisle Group had negative repercussions both psychologically and financially on the strikers. Prior to this, during the fall of 1921, letters from the Pulp & Sulphite strike organizers to President Burke indicated that the Paper Makers were weak in a number of locations and asked Burke to contact J. T. Carey, President of the Paper Makers, to send help. The organizers in the field could see the effect the breakdown in solidarity was having locally and warned Burke that the Paper Makers needed to get involved.

But the Paper Makers did little to help. An 11-year tradition of solidarity, created at a great price in human suffering, was destroyed by an ally in the heat of battle. Workers in both unions were demoralized and the cut in wages taken by the Pulp & Sulphite workers made it almost impossible for them to continue giving financial support to their brothers and sisters on strike against the International Paper Company. Meanwhile, this action by the Paper Makers Union and the Carlisle Group was seen as a good sign by the managers of the International Paper Company. The unity of the unions had been broken.

International Paper Company's Trump Card

International Paper Company has always had very deep pockets. The company invested heavily in the Three Rivers, Quebec mill it had bought in 1919, hoping the mill would be able to replace production at the struck mills. The company started to shift production to this mill in 1922; by 1926 it was the largest paper mill in the world.

The three blows absorbed by the Pulp & Sulphite — the first from the Carlisle Group; the second from the Paper Makers Union and the third from International Paper Company's Three Rivers mill took their toll. Although thousands of workers, their families and communities continued to fight, more and more workers crossed the picket lines. By the end of 1924, approximately 40 percent of the Pulp & Sulphite Union workers who struck in 1921 had returned to work in the mills of the International Paper

Company.[25] As time went on, more and more of the striking locals stopped functioning and the strike was officially called off on March 20, 1926. The five-year battle in the mill towns was over. The rank and file union that "perfected the organization" in 1910 was dead. A new union would emerge in 1934 in the wake of the general strike in San Francisco on the West coast and behind the organizing of the CIO in 1937 on the East coast.

Similar Times — Similar Strikes, 1921/1987

In 1987 workers in Jay and Livermore Falls were aware that the last time they struck International Paper Company was in 1921. Most of them knew the strike was lost, and many knew who some of the scabs were. Just before the strike of 1987/88 a union member in Jay informed me that the father of a local union officer had been a scab in 1921. I asked the brother who won the strike of 1921; he wasn't sure, he thought the union lost.

George Lambertson grew up in Berlin, New Hampshire and was about to retire as an International Representative of the UPIU when the strike of 1987 began in Jay. Early in the strike I asked George if he knew very much about the strike of 1921. He said he didn't but would ask his mother because she had lived through that strike. After he talked to his mother he related to me that the strike of 1921 to 1926 was a bad one and he recommended that we not talk about it at the mass meetings. One thing was clear in 1987; there was little historical memory of the strike of 1921, either locally or in the national union, beyond knowing who some of the scabs were and that the strike was lost.

There are many similarities between the strikes of 1921 to 1926 and 1987/88 and the times in which they occurred. Calvin Coolidge was President in 1923 and Ronald Reagan in 1987. As Governor of Massachusetts, Coolidge became well known when he broke a police strike in Boston by firing the entire force; Reagan set the tone for labor relations in the 1980s by permanently replacing all the striking air traffic controllers.* During both decades the rich enjoyed massive tax breaks and deregulation of corporate power was the order of the day. Conventional wisdom held that deregulation would create jobs. The "trickle down" theory prevailed in both decades. That is, the corporate class was trickling on the rest of us.

Both in 1921 and 1987, the company provoked the strikes and the national union seriously underestimated the company's financial strength and ideological commitment. In the final analysis, workers struck in both cases

---

* Ironically, in 1998, the U.S. Congress voted to name Washington, D.C.'s airport after Ronald Reagan, and the Boston Police Patrolmen's Association after an absence of 80 years joined the AFL-CIO.

because their backs were up against the wall, and win, lose, or draw they had to fight for their dignity. Many people had to relocate to find work. Families were divided, homes lost and meals missed. But most importantly, the dreams parents had that their children's lives would be better than their own were dashed in the aftermath of both strikes.

Ideology

The tyranny of International Paper Company was part of a nationwide corporate assault upon the U.S. working class. The tyrants weren't simply trying to make a better or more profitable product but to impose a "new world order." It is possible the International Paper Company managers were prepared to fight to the end for this new world order (the end for them being bankruptcy). A 1948 International Paper Corporation written and published

Paper Workers' Journal. January 1911.

book stated: "In 1921, the Company had a severe and crippling strike which brought both the Company and unions close to disaster." This might have happened if the unions had not cracked first.[26]

Key to understanding the union's defeat in the 1921 strike is the corporation's ideological opposition to unions. That ideology surfaced in Livermore Falls when the Chamber of Commerce invited International Paper Company to send a representative to a public meeting at which it was hoped both sides would present their positions on the strike. The company responded with a telegram reprinted in the *Lewiston Evening Journal* on July 22, 1921:

> You will take notice that we will not tolerate interference, that we will employ directly the men we desire and we will hold this town responsible for disturbances of peace and injury to persons and property. It remains to be seen whether the Chamber of Commerce stands for Bolshevism.

Needless to say, the company didn't send anyone to speak at the public meeting. (It never sent anyone to appear on a public platform with strikers in 1987/88, either.) The company made its position very clear: it would not only decide how the mill was going to be run, but how the *town* was going to be run. Furthermore, International Paper Company, in alliance with other major U.S. corporations, was going to decide how the *country* was going to be run.

DEMOCRATIC INSTITUTIONS CHALLENGED IN LIVERMORE FALLS

The idea that a democracy only works when there is a free and public exchange of ideas had no bearing on International Paper Company's position. Even when the Chamber of Commerce dared to sponsor a public debate on the most important issue facing the community, the company felt no need to participate. The company's position was that no one had the right to question it, and if they did, they must be Bolsheviks.

AND AROUND THE COUNTRY

- On January 2, 1920, U.S. Attorney General A. Mitchell Palmer launched the largest of the "Palmer Raids." His agents arrested more than 10,000 people for their alleged radical activities. "In Boston 400 workers were led through the streets manacled and handcuffed, clanking with chains." [27] Eventually, the majority were released but 3,500 were held for deportation hearings. In total close to 1,000 people were deported. The purpose of the raids was to suppress political activity by arresting foreign-born people active in unions and radical politics. Most factory workers at this time were immigrants and children of immigrants.

- The New York State Assembly in 1920 expelled five duly elected Socialist Party legislators.
- During 1921, 10,000 armed members of the United Mine Workers Union marched on company coal mining towns in West Virginia to liberate those communities from the control of the coal corporations. This was the largest movement of armed workers this country has ever seen. When they were finally turned back by Federal troops and defeated, the union went into a decline for the rest of the decade but came back fighting, and even organizing paper mill workers.
- Two of the most famous labor martyrs of all time, Nicola Sacco and Bartolomeo Vanzetti, a shoe maker and a fish peddler, were framed, put on trial in 1921, found guilty and sentenced to death in nearby Massachusetts.
- In April 1921, 46 organizers of the Industrial Workers of the World, the cutting edge of organized labor, were put in Federal prison and one of labor's best known leaders, "Big Bill" Haywood, went into exile rather than go to jail.

With most of its leaders in jail, exiled or deported, the I.W.W. never regained its position in the labor movement.

No Past, No Future

In 1987, Local 14 requested a history of the paperworkers unions' struggle with International Paper Company from their national union, the United Paperworkers International Union. They were sent a document one and one-half pages long, entitled *The United Paperworkers and International Paper: Ninety Years of Struggle.* The strikes of 1908, 1910, 1918 and 1919 were not mentioned and the full description of the strike of 1921 follows:

> International Paper, however, refused to submit to arbitration. The workers called a strike: injunctions were issued, troops called out, workers jailed and families evicted from their company-owned homes. Several strikers were shot by men employed by the company. In the end, 22 locals, with more than 9,000 members were wiped out of existence. After five bitter years, defeated workers returned to the IP mills.

Although many union workers returned to the mills of the International Paper Company, the unions didn't return until 1937.

It's impossible to know if the outcome of the strike of 1987/88 would have been any different if the workers who struck had known the history of the previous strikes. But had they known the history, they would have had more information on which to base their decision to strike, and they would have had more information with which to wage their strike. Why was this history buried?

ENDNOTES

1. "Strikers Hold Big Mass Meeting At Rumford." *Lewiston Sun.* August 11, 1921.
2. "I.P. Co. Asks Men to Return at 21percent Cut." *Lewiston Sun.* June 3, 1921.
3. "Labor Officials and I.P. Co. Deadlock at Conference." *Lewiston Sun.* June 21, 1921.
4. "Paper Mills May Reopen Next Week." *Lewiston Sun.* June 29, 1921.
5. "International Paper Co. to Open Plants July 5." *Lewiston Sun.* July 2, 1921.
6. "Livermore Falls Citizens Doing Their Utmost." *Lewiston Evening Journal.* July 22, 1921.
7. "Strikebreakers Are Attacked at Corinth." *Lewiston Sun.* July 22, 1921.
8. "Request That Strikebreakers Be Kept Out of E. Livermore." *Lewiston Sun.* July 22, 1921.
9. "Sharp Reply Sent to East Livermore." *Lewiston Sun.* July 23, 1921.
10. "Troops Mobilized for Possible Duty at Bellows Falls." *Lewiston Sun.* July 27, 1921.
11. "Rumors of Threats to Dynamite I.P. Co's No. Walpole Booms." *Lewiston Sun.* July 29, 1921.
12. "Labor Leaders Confer with Gov. Baxter Regarding Paper Mill Strike." *Lewiston Sun.* July 30, 1921.
13. "Tenants of I.P. Co's Houses Ordered to Move Out." *Lewiston Sun.* August 2, 1921.
14. "100 Strikebreakers Arrive at Livermore Falls." *Lewiston Sun.* August 6, 1921.
15. "International Begins the Operation of Sulphite Mill." *Lewiston Sun.* August 10, 1921.
16. "I.P. Co. Closes Rumford Plant for Six Months." *Lewiston Sun.* August 19, 1921.
17. "Report Strikers Damaging I.P. Camps in West Enfield." *Lewiston Sun.* August 26, 1921.
18. "Will Try to Settle the Paper Mill Strike Today." *Lewiston Sun.* September 1, 1921.
19. "International Paper Company's Relations with Its (sic)Employes." *Lewiston Sun.* September 20, 1921.
20. Voelker 196.
21. Gonya.
22. "22 Guilty of Conspiracy." *Lewiston Evening Journal.* November 12, 1921.
23. Voelker 210.
24. Voelker 210.
25. Voelker 215.
26. *International Paper Company After Fifty Years* 21-22.
27. Boyer 212.

THE CORPORATE REPUBLIC 9

When International Paper Company gained the upper hand by defeating the unions in the strike of 1921 to 1926, it made good on its promise to run the mills as open shops and institute the "American Plan." The unions' influence in the community was replaced by the influence of the company and the church.

Murray Hall

The union hall as a center of community life died, and in its place the company had a community hall of its own erected in Livermore Falls. The company hall, Murray Hall, was named after a local mill manager, W.A. "Bill" Murray, who had many relatives in town. Murray Hall is still a center of community activity to the present day. The union hall, with the exception of the 1987/88 strike years, never regained its place as a center of community activity as it was prior to the strike of 1921.

Strikers of 1987, who traveled around the country visiting workers who lived in International Paper Company mill towns, told me that in many of the towns they were amazed to find halls just like Murray Hall. Had they known the labor/corporate history of 1921 through 1937, they would not have been amazed. They would have been surprised *not* to find them.

EMBA

The company organized the Employees Mutual Benefit Association (EMBA) with a chapter at each mill to act as a financial safety net for workers, a role once fulfilled by local unions. So if your house burned down or you were out of work due to an extended illness, you went to the EMBA for help instead of the union. The local EMBA chapters were made up of management and hourly people. In the modern era the EMBA received its money from a rake-off on the vending machines, and a majority of the EMBA board in Jay was made up of union members. (Shortly after the strike of 1987 began the EMBA Board voted to turn the entire bank account of more than $100,000 over to the strikers' food bank. The International Paper Company contested the Board's action in court, but lost.)

THE CHURCH

The Roman Catholic Church moved to the forefront as a force in the community to take the place of the defeated unions. Anthropologist August Carbonella, examining the relationship between the company and the Catholic Church in Jay after the defeat of the unions and the institution of the American Plan, writes:

> There were sound reasons for Magnan's (local mill manager) involvement in the Church.* With the collapse of the unions, St. Rose of Lima became the principal institution of social life in the French-Canadian community. Even if the clergy was not simply doing the bidding of mill management, their cooperation at least stemmed from an elective affinity of agendas. As Magnan's heavy involvement with the Church demonstrates, management undoubtedly equated church membership with the development of a stable and dependable work force. Priests, on the other hand, likely perceived the maintenance of a stable work-force as vital to community integration and the health of the parish. The frequent priestly sermons on the importance of sobriety, family responsibility, and pre-marital chastity very likely cheered Magnan considerably as he cast his gaze on the congregation from his pew in the rear.[1]

The unions had previously held the high ground over the Church *because* they advanced the idea that heaven should be created on earth. In contrast, church leaders told people if you quietly accept your place on earth without protest, you will be rewarded and go to heaven when you die. Joe Hill, I.W.W. organizer, dreamer and writer of proletarian songs, who was framed and hanged in response to an I.W.W. organizing drive in Salt Lake City, reflected on the relationship between the church, the company and the union in a song (circa 1911):

* According to Carbonella, mill manager T. G. Magnan, although he wasn't Catholic, attended Catholic services every Sunday.

THE PREACHER AND THE SLAVE

(Pie in the Sky)

Long-haired preachers come out every night
Try to tell you what's wrong & what's right
But when I asked 'How bout something to eat?'
They will answer in a voice so sweet:

You will eat, bye & bye
In that glorious land above the sky
Work & pray, live on hay
You'll get pie in the sky when you die (That's a lie!)

Working people of all countries, unite
Side by side we for freedom will fight
When the world & its wealth we have gained
To the grafters we will sing this refrain:

You will eat bye & bye
When you learn how to cook & to fry
Chop some wood, 'twill do you good
And you'll eat in the sweet bye & bye (That's no lie!)

ENDNOTE

1. Carbonella 122.

Union Membership by Year in Unions Nationally and in the Paper Industry 1936-1961			
Year	**Paper Makers**	**Pulp & Sulphite**	**All Unions**
1936	7,700	17,000	4,100,000
1941	36,900	60,000	8,700,000
1946	51,200	97,700	13,000,000
1951	67,600	141,600	15,100,000
1956	80,300	165,800	16,500,000
1961	126,200	171,100	15,400,000

Note: Data for different unions are not strictly comparable—provided for trends only.

10

THE CREATION OF A BUREAUCRATS' UNION: ORGANIZING BUSINESS LEADERS

The Pulp & Sulphite Union grew from fewer than 6,000 members in 1933 to more than 120,000 in 1948. According to Robert H. Zieger:

> *Three central factors explain the Brotherhood's resurgence in the 1930s. The first is the sheer fact of the union's survival through years of defeat and decline — its ability to cling to its broad jurisdiction and to remain capable of exploiting opportunities to extend its organization. The second is labor legislation passed by Congress, notably Section 7(a) of the National Industrial Recovery Act (NIRA) and the National Labor Relations Act of 1935. The third is the challenge presented by the creation and activities of the Congress of Industrial Organization. Both workers and employers were central. The workers themselves, of course, had to be willing to support the union. Influential employers also helped make the climate favorable for unionism. In some places, arrogant and hostile employers created situations that the union eagerly seized upon; elsewhere, corporate executives openly welcomed members of the Brotherhood, viewing acceptance either as an adjunct to modernize personnel policies or as a hedge against the aggressive CIO.*[1]

The unions that were beaten in the strike of 1921 returned to the mills in 1937, but by then these unions were very different from the ones people had known from 1910 through 1921. The new unions were built on a very different premise, with different goals, strategies and tactics. The leadership did not want to go back to the old ways; the years of struggle and defeat had softened them up and led them down a new path. They didn't want the new membership to see the old ways as an alternative to the present, so the institution changed and the history was buried.

FRANKLIN ROOSEVELT

The country seemed to prosper under *laissez-faire* capitalism during most of the 1920s until 1929. Then the stock market crashed, banks failed and the country plunged headlong into the great depression. The voters lost confidence in the deregulation and trickle-down economics that a decade of Republican rule had brought, and in 1932 Democrat Franklin Roosevelt was elected President.

Roosevelt was a liberal capitalist. His administration took the view that the way to stimulate the depressed economy was *not* by cutting wages, taxes and regulations and destroying unions (all of which had been done in the past decade), but by setting up regulatory agencies, raising taxes and encouraging the formation of unions. The unions would then fight for better wages. Better wages would mean workers would spend more money to buy things and more jobs would be created to satisfy the needs of workers spending more and more money. The opposite of trickle-down: trickle-up. Trickle-down argues that the more money you put in the hands of people at the top, the more money they have to spend and invest. The more money the rich spend and invest, the more jobs are created and then the economy starts to hum. A major difference between

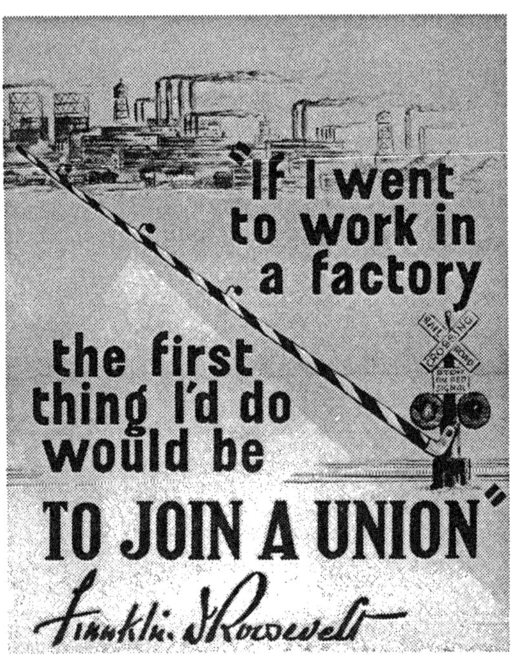

CIO Recruiting Poster

trickle-down and trickle-up is that the rich will buy only so many luxury yachts and save the rest, but there is no limit to the number of cars workers will buy if they have the money. Thus, every dollar the working class gets is spent, thereby stimulating the economy.

Activist Government / Regulated Unions

Roosevelt Democrats, unlike the Republicans, believed the government should take an activist role in regulating financial institutions to prevent bank failures and stock market crashes. They also believed unions could be a force to stimulate the economy. But they did not want the destabilizing effects that wildcat strikes and general strikes had on the economy when labor and capital were engaged in struggle, nor did they want the ideological polarization. That is, they did not want labor to challenge any of the basic underpinnings of corporate capitalism but they did want unions to stimulate the economy. The solution was to have unions that were defined and regulated by government. Defined narrowly by the government so they would not create an ideological polarization and regulated so they wouldn't engage in wildcat strikes and strikes that threatened "national security." So when the National Labor Relations Act was written, its constitutional underpinning was the Commerce Clause of the Constitution — not the 13th Amendment which outlaws conditions of involuntary servitude or the 1st Amendment which guarantees freedom of speech, assembly and association — rights labor had been fighting for since the end of the Civil War.

Safety Net

Under Roosevelt, the first Federal "safety net" was set up to give a pension (called Social Security) to workers too old or too sick to work. A system of Federal/State Unemployment Insurance was created so laid-off workers would have a small income to tide them over while they looked for work. Another Federal/State program called Aid to Families with Dependent Children (AFDC) was instituted to protect children with one parent who had little or no income. The Roosevelt Administration also set up the Federal Deposit Insurance Corporation (FDIC) to insure the savings accounts of working class people in case of bank failures, which had become all too common.

High Unemployment / Low Wages

In 1933, one-third of the work force was unemployed and many of the employed were working part time. For those lucky enough to be working, real wages had fallen 16 percent in two years.[2]

Roosevelt's programs did a lot to make the lives of average people better and to give people hope, but it took World War II and deficit spending to get the nation out of the depression.

Government and Unions Reconsidered

The stock market crash of 1929 and the depression that followed forced the nation to reconsider the role of government in solving economic problems. Likewise, the loss of the strike of 1921 through 1926 was the event that pushed the paperworkers union leadership to reconsider the role of the membership in the union and the relationship between the union leadership and the corporations.

National Industrial Recovery Act and Militant Unions

Passage of the National Industrial Recovery Act (NIRA) in 1933 and the rise of militant unions were the two key factors that pushed the paper companies into the arms of the Paper Makers Union and the Pulp & Sulphite Union. This marriage transformed both organizations from member-based unions to employer-based unions. Through the strike of 1921 the unions relied on the activity and solidarity of the membership to force recognition and benefits from employers. After passage of the NIRA in 1933, the AFL paperworker unions relied primarily on government regulations and the threat of militant unions organizing paper mill workers to bring about recognition and benefits from employers.

The purpose of the NIRA was to stimulate the economy by increasing production and therefore employment as well. Some of the programs of the NIRA included the appropriation of large sums of money for public works projects and the creation of a process, governed by the National Recovery Administration (NRA), to develop codes that would govern almost every aspect of industrial activities to "end cut throat competition, eliminate overproduction, increase industrial efficiency, and raise prices, profits, and wages."[3] These codes were written by the corporations and their trade groups with the help of government, consumer and labor leaders.

Workers Exchange Fear for Hope

President Roosevelt encouraged workers to join unions. And he didn't just "talk the talk, he walked the walk." Section 7a of the NIRA says "employees shall have the right to organize and bargain collectively through representatives of their own choosing . . . free from the interference, restraint, or coercion of employers."

The onset of the depression, in which one in three were desperate for any work, made most employed workers fearful to take any job actions in the late 1920s and early 1930s. But the combination of the corporations' cutting wages and increasing work loads and the administration's supporting workers' rights helped reverse the climate created by political and economic suppression which characterized the decade of the 1920s.

National Unions not Prepared

The dam broke as millions of workers exchanged their fears for hope and started to organize. But the national unions were caught unprepared. With the I.W.W. now out of the picture and the surviving AFL unions no more prepared to organize the industrial workplace of the 1930s than they were to organize the steel industry in 1919, workers seeking to organize were often turned away by unions.

> For instance, leaders of the Pulp and Paperworkers' Union, an old line AFL union, did not launch an all-out organizing crusade. Workers flocked to the union anyway. 'I have so many calls for organizers,' the union's president declared, 'that I have neither the men nor the money to take care of them.' [4]

Even though the leadership was unprepared for the change, the national membership of unions doubled during Roosevelt's first term. But the presidents of the two paperworkers' unions had put their eggs in another basket. They were going to get the corporations to help them organize and they were sure the workers would follow.

The Stage is Set

President John Burke of the Pulp & Sulphite Union and President Matthew Burns of the Paper Makers Union were appointed advisors to the federal administrator of the paper industry codes. The hearings, meetings and activities that the creation of the codes presented gave Burke and Burns many opportunities to meet with the paper industry leaders. Burke reported on August 29, 1933,

> There must have been about four hundred manufacturers at this hearing, practically all open shoppers. I think the hearing has broken down a great deal of opposition to our unions, because both Burns and I made a good impression at the hearing. If we use our heads now and do not make wild demands, I feel that we have an opportunity to organize the whole industry.[5]

Organizing in this fashion, from the top down, went well for Burke and Burns. Their first major success was on the West Coast. There the entire industry, which had been unorganized in the late 1920s, was organized in 1934. The success on the West Coast was followed on the East Coast in 1937 when the International Paper Company recognized the paper worker unions. These were the same unions that the International Paper Company drove out of the mills in 1921. But now they were acceptable. Maybe they weren't really the same unions?

Organizing on the West Coast

President Burke began the process of organizing on the West Coast by first identifying an industry executive who was favorable to unions. Next he wrote to AFL President William Green on June 27, 1933, explaining that the climate on the West Coast was now favorable to organizing in the pulp and paper mills.[6] He asked Green to assign people to start organizing locals for the paper worker unions. Green responded and sent an AFL organizer who set up several locals, and in January 1934, the locals formed The Employees Association. These locals were very small, and not recognized by the companies, but provided the national unions, headed by Burke and Burns, with a foot in the door. Burke and Burns followed this up by sending George Schneider out to the West Coast. Schneider was the Paper Makers Union organizer who was run out of Rumford, Maine, by vigilantes during the strike of 1910 and was now a US Congressman from Wisconsin. Schneider spent some time with the new locals and much of his time with industry people.

Herbert Sullivan followed Schneider as a West Coast organizer. Sullivan was appointed a vice-president of the Pulp Workers Union by Burke and assigned to head up the work. Sullivan wrote Burke on July 13, 1934, that all the employers had agreed to meet with the unions.[7] On August 1, 1934, The Employees Association, with Burke at the helm, sat down to bargain with the employers. Out of the negotiations a Uniform Labor Agreement (ULA) was established in which the Pulp & Sulphite Union and the Paper Makers Union were recognized by 14 employers under Section 7a of the NIRA.[8] Wages were set at 45 cents an hour for men and 37 cents an hour for women. Three holidays were granted, a grievance system terminating in a Joint Relations Board was set up, and a joint committee was established to determine the prevailing wage for various jobs. The entire work force on the West Coast in the primary pulp and paper industry was now organized from the top down. There were no strikes for recognition, no elections, little if any rank and file organizing or organization. Something like instant unionism:

Just add a depression, the New Deal and leave it to someone else to do the real work.

Others Do the Real Organizing

It is very important to point out that the recognition of the unions by the employers on the West Coast did not occur in a vacuum. In fact, this may be an understatement. Consider the following events that were unfolding while the paper industry was sitting down to bargain with the paperworker unions on the West Coast. The Longshoremen struck the docks from Puget Sound to San Diego on May 9, 1934. The Teamsters boycotted the docks and the maritime unions also struck. The police, on July 5, while trying to open the port of San Francisco for business, fired on a crowd, killing two people. A general strike was called on July 16 and throughout the city of San Francisco 130,000 workers just stopped working. Newspaper columnist Mike Quin described it this way,

> Highways leading into the city bristled with picket lines. Nothing moved except 'By permission of the strike committee.' To all intents and purposes, industry was at a complete standstill. The great factories were empty and deserted. No streetcars were running. Virtually all stores were closed. The giant apparatus of commerce was a lifeless, helpless hulk. Labor had withdrawn its hand. The workers had drained out of the plants and shops like life-blood, leaving only a silent framework embodying millions of dollars worth of invested capital. In the absence of labor, giant machinery loomed as so much idle junk. ... Labor held the life-blood and energy. The owners remained in possession of the corpse.
>
> An atmosphere of panic prevailed in the business districts. But an almost carnival spirit was apparent in working-class neighborhoods. Laboring men appeared on the streets in their Sunday clothes, shiny celluloid union buttons glistening on every coat lapel. Strangers addressed each other warmly as old friends. Labor wore its new-found power with calm dignity. After all, this great intricate mass of steel, stone, and machinery was their baby. They built it and they ran it. They knew every bolt and cog and screw in the apparatus of civilization. They could turn the whole thing off at will, or start it up again if they pleased. They were organized.[9]

The Longshoremen's strike was settled through arbitration and they went back to work at the end of July.

The Longshoremen's strike affected the paper industry in two important ways. First, the industry had no way to obtain raw materials or ship finished products through the ports because the ports were shut down. Secondly, the power the unions were demonstrating had to weigh heavily on the minds of paper industry executives as they sat down to negotiate with the leaders of the paperworker unions. The employers had to be concerned that the militancy of workers and their organizations at the ports might spread to the paper mill towns.

The employers of the paper industry could not open the ports, but by recognizing the paperworker unions they hoped these now docile unions would help them keep the militant unions out of their mills. The industry was right. They got relatively docile unions. And the unions they recognized got a new lease on life: dues-paying members.

The AFL and the CIO

Not all unions were like the paperworker unions, waiting around for the companies to help them organize. Some, like the United Mine Workers led by John L. Lewis, were willing to continue their militant tradition of organizing. The Mine Workers were members of the AFL and pushed within the Federation to come up with ways to take in the millions of industrial workers who were clamoring for unions. The AFL's strategy of organizing workers in skilled trades didn't leave any room for the masses of workers now employed in industries like auto, steel and rubber.

At the AFL's convention of 1935, Charles Howard, president of the Typographical Union, said: "The workers of this country are going to organize, and if they are not permitted to organize under the banner of the American Federation of Labor they are going to organize under some other leadership."[10] But Howard's observation was received coldly at the convention except by a small group led by Lewis who said,

> Great combinations of capital have assembled to themselves tremendous power and influence, and they are almost 100 percent effective in opposing.. .. .. the American Federation of Labor.. .. .. . If you go in there with your craft union they will mow you down like Italian machine guns mow down Ethiopians in the war now going on in that country.[11]

Nothing productive came out of the convention, so Lewis called together a group of AFL union leaders who were sympathetic to organizing industrial workers by industry rather than by craft. They formed the Committee for

Industrial Organization within the AFL. Its express purpose was to promote organizing of workers in mass production industries.

The AFL demanded the Committee be scrapped but the founders refused, so the AFL expelled the leading unions; among them the Mine Workers,

UMW. *John L. Lewis and the International Union United Mine Workers of America; The Story from 1917 to 1952.* 1952.

Clothing Workers, Textile Workers and Mine, Mill and Smelter Workers. The Committee then changed its name to the Congress of Industrial Organizations, commonly known as the CIO, which was soon joined by new unions like the United Electrical Workers (UE) and the United Auto Workers (UAW). The unions of the CIO soon gained a militant reputation by leading numerous massive sit-down strikes.

Lessons of the CIO[12]

There are many in today's unions who would like to recreate the vibrant, militant and successful organizing drives of the CIO unions in the 1930s and 1940s. The new CIO unions were a product of a particular historic moment that will never be recaptured. However, certain elements had to be present to turn the social turmoil of the period into a union upsurge. Understanding these elements and how they contributed to the overall experience of the working class during that period may help us take advantage of the times we live in.

Six elements follow that played a role in the CIO's advance.

1. **A suprisingly small number** of dedicated, militant activists comprised of experienced organizers, linked politically and geographically, worked in the factories and formed cadres of activists committed to social change. Workers generally were extremely cautious and uncertain so organizational breakthroughs were dependent on this core of dedicated activists determined to build new unions.

   Many of these worker activists had survived the labor defeats that occurred after WWI and had continued to organize and learn from their ongoing experiences. These cadres implemented the breakthrough strategy of the sit-down strikes and mass picketing that characterized the initial organizing drives of the CIO.

2. **Once successful, union organization spread rapidly** through plants of the same corporate chain and geographically in the cities where the sit-downs were taking place.

3. **National leadership and resources** were provided by a few top union leaders who broke politically and tactically with the AFL. They refused to compromise militancy where it developed and provided local organizing efforts with a national voice and support.

4. **CIO leaders and organizers** united workers politically in key industrial states and delivered the vote in the re-election of Roosevelt in 1936 and the election of New Deal governors in Michigan, Ohio, Indiana and Pennsylvania.

   These electoral successes increased working class consciousness and confidence, emboldened union organizers' plans, and created a brief six-month window in which the sit-downers established new unions.

It was in this same, short-lived, political climate that the Supreme Court broke with its historic position that laws protecting labor rights were unconstitutional by finding the National Labor Relations Act constitutional.

5. **New Federal Laws** made possible a second round of organizing by the CIO and the AFL. The first round successes came through sit-downs; the second round through National Labor Relations Board recognition elections.

UMW. *John L. Lewis and the International Union United Mine Workers of America; The Story from 1917 to 1952.* 1952.

(Previous to the Norris-LaGuardia "Anti-Injunction" Act of 1932, strikes were often enjoined — stopped — by courts. Prior to the National Labor Relations Act of 1935, there was no federal mechanism, with the exception of the Railway Labor Act of 1926, to require employers to recognize unions.* In the beginning the National Labor Relations Board aggressively enforced workers' rights by reinstating thousands of discharged workers fired for union activity, forcing employers to recognize unions, and until 1941 made sure the corporate voice was silent in union recognition elections.[13])

6. **The new unions directly challenged management for shop floor power.** They utilized extensive stewards' systems with the direct involvement of the rank-and-file to settle grievances through job actions. This direct involvement of workers on the shop floor built meaningful in-plant organization which helped workers and the new CIO unions weather the significant corporate counter attacks that often followed union recognition.

These elements are intertwined. Each was set in motion by earlier decisions of union leaders, organizers and workers. They all combined to create the successes of the CIO. United Mine Workers President John L. Lewis recognized that workers in the key industries of steel and auto, dominated by huge corporations, could not be organized without the support of the President and governors of the states involved. This recognition led to the key decision to move CIO organizers into the political field for Franklin Roosevelt's race against the "economic royalists" in 1936. With a Roosevelt victory in hand, Lewis was able to state at a national press conference during the middle of the Flint sit-down strike early in 1937:

> We have advised the administration . . . that for the six months the economic royalists represented by General Motors contributed their money and used their energy to drive this administration out of power.
>
> The administration asked labor for help to repel this attack, and labor gave its help. The same economic royalists now have their fangs in labor. The workers of this country expect this administration to help the workers in every legal way, and to support the auto workers in the General Motors plants.[14]

The Governor of Michigan didn't unleash the militia and Roosevelt didn't send Federal troops. Instead, they pressured the corporate managers and owners to come to terms with the CIO. Six months later the CIO had almost 2 million members and industrial unions were a force to be reckoned

* "Judges in both state and federal courts issued more than 1,800 injunctions against strikes between 1880 and 1931." Fink 150.

with.* The managers of the International Paper Company could see the writing on the wall. So rather than wait for the CIO to come calling, they made a deal with their old adversaries the leaders of the Paper Makers and Pulp & Sulphite unions.

Back in Jay the Company Helps the Union

While the unions of the CIO were locking companies out with sit-down strikes, Mickey Poulin went to work for the International Paper Company in Jay, Maine. According to Mickey, in 1936 most of the paperworkers in Jay were scabs from the strike of 1921. One day in 1937 Mickey's foreman told him if Mickey wanted to continue in the employ of International Paper Company he would have to join the union that night and bring a union card to work the next day.[15] The third paragraph of Section 2 of the contract between the International Paper Company and the Paper Makers and Pulp & Sulphite Unions at the Otis Division in Livermore Falls, Maine; the Hudson River Mill in Palmer, New York; the Fort Edwards Mill in Fort Edwards, New York; and the Niagara Mill in Niagara Falls, New York, signed on October 28, 1937, states: "The International Paper Company through its local management will cooperate with the local unions in every way the management of the Company considers proper and lawful to assist in obtaining and retaining members." That is how the paperworkers' unions came into the life of many Northeastern paperworkers in 1937. Not through company-contested NLRB elections, local community pressure or sit-down strikes, but through management/union cooperation!

One, Two, Three: Canada, Northern Division, Southern Kraft

International Paper Company, then as now, was the largest paper company in the country. According to Zieger,

> Throughout 1937, the paper unions and International Paper officials developed an approach to organizing the company's workers. In April the parties met in Montreal and signed union shop agreements covering the operations of the Canadian International Paper Company, whose mills produced vast quantities of newsprint. The normally phlegmatic Burke was ecstatic about this development. 'These negotiations were of such importance,' he observed '.. . . because from May 1, 1921 to the signing of this agreement . . . all of the mills of the International Paper Company have been open shop.'[16]

* According to Troy 8. The CIO claimed 3,419,600. Dubofsky and Van Tine 278.

The Canadian agreement was followed by agreements with International Paper Company in the northern U.S. and shortly thereafter in the South with International Paper Company's southern division known as the Southern Kraft Paper Company. The new contracts were reflected in the growth of the Pulp & Sulphite Union which went from a membership of 5,684 in 1933 to 45,000 in 1938 and by 1948 the union had 123,664 dues paying members.[17]

FACTORS

Similar factors pushed International Paper Company management to sign agreements with the paperworkers' unions on the East Coast as on the West. The Federal government's influence was the same on both coasts through the NIRA in 1933 and then with the NLRA (National Labor Relations Act) of 1935. In the East, the CIO Paper Mill Workers Union had started to have some success in organizing paper mill workers.* International Paper Company managers, remembering the strike of 1921, wanted no part of John Lewis and the CIO. They didn't want sit downs in any paper mill towns so they signed union shop agreements with the Paper Makers Union and the Pulp & Sulphite Union.

Similarly in the steel industry it was the organizing success that the auto workers had with its sit downs at General Motors in early 1937 that pushed the President of U.S. Steel, the second largest manufacturing company in the world at the time, to sit down with CIO chief John L. Lewis. Out of their meetings came a contract signed on March 2, 1937 which included recognition of the Steel Workers Organizing Committee as bargaining agent, a grievance procedure, 5 percent wage increase, 40-hour week and time and one-half for overtime.

LESSONS

Meanwhile, the lessons that the leadership of the paperworkers' unions learned were to go with the flow, take the path of least resistance and preserve the bureaucracy. They wanted to avoid at all costs a community-based, rank-and-file, active industry wide, worker-based union as had existed from 1910 to 1921. Trade it all for company recognition with government protection and regulation.

---

* Later the Mine Workers set up District 50 which won bargaining rights at the following mills of the: Oxford Paper Company in Rumford, Maine; West Virginia Pulp and Paper mill in Mechanicsville, New York; Kimberly-Clark in Niagra Falls, New York; Scott Paper Company in Fort Edward, New York (Graham 22).

The leadership learned to let somebody else do the work for them. They turned their backs on the union that based its power on community support; the type of union that once played a part in every aspect of working class life. The leaders traded in that union for a sterile union that relied on the Federal government and the threat of militant unions for its existence. And President Burke of the Pulp & Sulfite Union who in 1908 said, ".. .. .. every stand of the working class against the criminal greed and avarice of the capitalist class is a step forward in the aspiring march of the masses toward their final and complete emancipation." was saying in 1929, ".. .. .. we do not propose any radical changes but are willing to work along slowly and assist the Company in making the mill a success."[18]

Decline

As the years went by and the massive organizing drives of the CIO faded into the past, more and more unions fell into the same trap the paperworkers had. But this time, the only power the unions had to fall back on was the power of the Federal government through the National Labor Relations Board. So in 1981, when President Reagan fired the air traffic controllers (Professional Air Traffic Controllers Organization — PATCO) for striking, he reversed the signal that President Roosevelt had given 48 years before when he signed the NIRA and section 7a became law. Franklin Roosevelt encouraged workers to join unions. Ronald Reagan encouraged corporate managers to destroy unions.

Most unions by 1980 had lost the knowledge to organize and with it went the ability to respond to the corporate offensive. Unions, like muscles that don't exercise, had gotten weak and lazy hoping that government power would sustain them. The Carpenters/*Corporations* took the Oysters/*Unions* for a 50-year walk down the beach.

'O Oysters, come and walk with us!'
  The Walrus did beseech.
'A pleasant walk, a pleasant talk,
  Along the briny beach:
We cannot do with more than four,
  To give a hand to each.'
The eldest Oyster looked at him,
  But never a word he said:
The eldest Oyster winked his eye,
  And shook his heavy head —

Meaning to say he did not choose
To leave the oyster-bed.
But four young Oysters hurried up,
All eager for the treat:
Their coats were brushed, their faces washed,
Their shoes were clean and neat —
And this was odd, because, you know,
They hadn't any feet.
'O Oysters,' said the carpenter,
'You've had a pleasant run!
Shall we be trotting home again?'
But answer came there none —
And this was scarcely odd, because
They'd eaten every one.

(From Lewis Carroll's, *Through the Looking-glass)*

If the Pulp Workers had consulted the old oyster they might have continued the work of perfecting the organization, as they did in 1910. Instead they went for a walk with the walrus and started increasing their membership with the help of employers on the West Coast in 1934 and on the East Coast in 1937. Might things have turned out differently if they had listened to the old oyster instead of the walrus? We will never know.

But if we are steeped in the many facets of labor history, at least we can make informed choices. For example, it was the leadership of the AFL that opposed the National Labor Relations Act on the principled grounds that the Act would define unions. The AFL wanted union leaders to define unions, not the government.

On another point against the NLRA, AFL President William Green, said in 1934:

> Elections in individual plants supported by the National Labor Board should not be confused with real collective bargaining .. .. .. there is a need for organizations of workers on a basis at least as wide as the organization of employers .. .. .. In the long run we must look to independent organizations of workers on a national or international basis for real collective bargaining.[19]

Could the AFL have been right? Is there something fundamentally wrong with the National Labor Relations Act? Was there another road that could have been taken?

Endnotes

1. Zieger 11.

2. *Who Built America.* Vol. II. 319.

3. *Who Built America.* Vol. II. 353.

4. *Who Built America.* Vol. II. 357.

5. Graham 15.

6. Graham 26.

7. Graham 30.

8. Employers: Weyerhaeuser Timber Co., Longview Fiber Co., Pacific Strawboard and Paper Co., St. Helens Pulp and Paper Co., National Paper Products Co., Columbia River Paper Mills Co., Oregon Pulp and Paper Co., Hawley Pulp and Paper Co., Crown-Willamette Paper Co., Rainier Pulp and Paper Co., Washington Pulp and Paper Corp., Olympic Forest Products Co., Fiberboard Products Inc., and Grays Harbor Pulp and Paper Co.

9. Quin 147-148.

10. *Who Built America.* Vol. II. 378.

11. *Who Built America.* Vol. II. 379.

12. Lessons of the CIO was taken from notes Ed Bruno used in presentations to Labor Party meetings on Workers' Rights that took place in 2001 and 2002.

13. NLRB v Virginia Electric & Power (1941).

14. *New York Times.* January 22, 1937. 6.

15. Poulin.

16. Zieger 145.

17. Gross 197.

18. Voelker 2.

19. Green, William.

Union Membership by Year in Unions Nationally and in the Paper Industry 1962-2001

Year	Paper Makers	Pulp & Sulphite	All Unions
1962	130,600	174,100	16,900,000
1972	[Paper Makers and Pulp & Sulphite merge to form the United Paperworkers International Union (UPIU)]		
Year	**UPIU**	**All Unions**	
1974	335,100	22,200,000	
1985	232,000	17,000,000	
1991	202,000	16,600,000	
1995	233,000	16,400,000	
1997	226,000	16,100,000	
1999	[Oil, Chemical and Atomic Workers Union and the UPIU merged to form Paper, Allied Industrial, Chemical and Energy Union (PACE)]		
Year	**PACE**	**All Unions**	
1999	293,000	16,500,000	
2001	274,000	16,400,000	

Note: Data for different unions are not strictly comparable—provided for trends only.

11

NATIONAL LABOR RELATIONS ACT

Purpose of the National Labor Relations Act

It is in the national interest of the United States to maintain full production in its economy. Industrial strife among employees, employers, and labor organizations interferes with full production and is contrary to our national interest. Experience has shown that labor disputes can be lessened if the parties involved recognize the legitimate rights of each other. To establish these rights under law, Congress enacted the National Labor Relations Act. (taken from a guide to the National Labor Relations Act prepared in the Office of the General Counsel, NATIONAL LABOR RELATIONS BOARD, circa 1980)

General Motors Corporation President Charles E. Wilson in testimony to the Senate Armed Services Committee, 1952 said: "What is good for the country is good for General Motors, and what's good for General Motors is good for the country."

SIT-DOWN STRIKE

In 1937 when the General Motors Company refused to recognize the United Auto Workers (UAW) in Flint, Michigan, the union employed a tactic called the sit-down strike. When workers strike using the tactic of the sit-down they don't leave the plant. They *occupy* it. This act, of course, violates the laws protecting corporate property. It is a lock-out, only this time the company, the property owner, is locked-out by wage labor, the employees. A sit-down strike is a reverse lock-out, a no-holds-barred struggle, pitting the power of workers and their allies against the power of a corporation and its allies.

In Flint, the workers were strong enough to repel the power of the company goons, local police and sheriffs. Michigan's newly elected Governor Frank Murphy was pro-labor, Franklin Roosevelt was in the White House, and neither was prepared to send Government troops into battle against workers seeking union recognition. After 44 days, General Motors called a truce and agreed to recognize the United Auto Workers Union for six months at 17 plants. Auto workers at Chrysler Corporation followed suit and the membership of the UAW grew from 30,000 to 195,000 in less than a year. Back in Flint, workers elected pro-union candidates to local offices and took control of the local government away from the General Motors Corporation.

It was the power labor demonstrated through the sit-down strikes that created the momentum to force some major corporations to acquiesce and utilize the National Labor Relations Board (NLRB) over the next three decades.

In order to get a jackass's attention, it is said you need to first whack it with a two-by-four. The sit-down strike was labor's two-by-four.

Management Goes on the Political Offensive

The National Labor Relations Act (NLRA) was found constitutional by the Supreme Court in *NLRB v Jones & Laughlin Steel Corp* on April 12, 1937. As the NLRB began to take hold as the definer of union activity and as the arbitrator and jury of relations between labor and industry, management went on the offensive in the courts, Congress and the media. First the Supreme Court in *NLRB v Mackay Radio* (1938) establishes the principle that employers have the right to permanently replace strikers. Then the

Court in *NLRB v Virginia Electric & Power Co.* (1941) decided that employers could present their side in union certification campaigns even though the company is not on the ballot. With these cases the Court made it harder to strike and harder to organize.

Several laws were passed which considerably weakened the National Labor Relations Act and unions in general. Most notable was the Taft-Hartley Act of 1947 which:

- forced foremen out of the unions
- allowed state legislatures to ban the union shop
- outlawed the closed shop
- made sympathy strikes and secondary boycotts illegal for all practical purposes
- barred from participating in NLRB elections unions that didn't ban Communists from their membership
- took control of pension funds, and health and welfare funds away from unions
- allowed employers the right to actively and vocally oppose having labor unions in their enterprises
- created the Taft-Hartley injunction whereby the President can set in motion injunctions against "national emergency strikes" that "imperil the national health or safety"

Conservative and opportunistic labor leaders used the banning of Communists and the anti-Communist tenor of the times to banish almost all the progressive and radical labor leaders, thinkers and organizers from union leadership positions.

However, not all unions took Taft-Hartley lying down. On September 1, 1948 Local 299 of the Pulp & Sulphite took out the advertisement to the right in the *New York Times.*

Sadly, not enough of organized labor had the fighting spirit of Local 299 and the

> # THANKS!
>
> **MR. TAFT and MR. HARTLEY**
>
> **NOT for what you tried to do to us and to all the workers in this country. But thanks for waking us up . . . because that is precisely what your iniquitous law did.**
>
> ---
>
> We, of Organized Labor, are therefore, actively participating in the unceasing struggle to repeal your **UNFAIR** and **UNDEMOCRATIC** legislation. On this coming Labor Day we pledge ourselves to do all in our power to bring about repeal of the Taft-Hartley Law.

Paid Advertisement. *New York Times.* September 1, 1948.

seeds of labor's long decline in the second half of the twentieth century were allowed to germinate.

Union Recognition Elections: A Participant's View

The political climate for workers and unions had gone quite far down the chute when I had my first experience in 1976 with organized labor and the National Labor Relations Board. As a worker in a large shoe factory, I participated in a campaign for union recognition. We lost the election and I was fired. Finally, after two years my case was to be heard by an administrative law judge. But the company was in the process of closing the plant and moving the operation overseas, so the union organizing committee recommended that I take a deal offered by the company. The most important lesson for me was that the company could fire the people active in organizing a union and keep them out of the plant long enough to ensure that a strong union was never built.

Later I was elected president of a shoe workers local union at another shoe shop, and when that shop closed I joined the Painters' Union and was active as a shop steward.[1] I have also worked for and with several national unions and as the field representative of the Maine AFL-CIO. In the past quarter of a century I have experienced first-hand the workings of the National Labor Relations Board from its role in the certification process, bargaining process, decertification process and all the steps, appeals and just plain bullshit in between.

Justice for workers? Not through the NLRB!

My participation and observations have led me to conclude that workers cannot obtain justice through a quasi-judicial regulatory agency that has as its bottom line, "It is in the national interest of the United States to maintain full production in its economy." While it may generally be true that a full production economy is in the interest of the working class, it is true only if workers have justice. Justice in this case means that workers define what a full-production economy is and what it produces, and share with each other the wealth that a full-production economy creates. However, one should not assume because one lives in a society that has a full-production economy, workers' rights will be maintained or respected. As Jesse Jackson said in a speech he gave in Jay during the strike of 1987/88: "Let's remember, under slavery we *all* had jobs."

Good Faith Bargaining?

In general, under the National Labor Relations Act, good-faith bargaining is said to exist as long as both sides will sit down at the same table. That is the

Board's working definition of good-faith bargaining, so let's see how it played out in Mobile, Alabama.

International Paper Company, while making record profits in 1987, wanted to get work rule and monetary concessions from its union employees in Mobile.* The company said it needed concessions to be competitive. The union said the company didn't need the monetary concessions because it was making huge profits. After the local unions voted to reject the company's proposal, the company locked the workers out. That is, they said if you won't accept our last offer we will not allow you to work and we will hire temporary workers to do your jobs until you accept our contract offer. After being on the streets for 19 months, the workers agreed to sign a contract that was only marginally better than the one previously offered and worse than their previous contract.**

Three months after International Paper Company began the Mobile lockout, workers at three other locations went through similar negotiations with International Paper Company and decided to strike against the proposal and in support of their sisters and brothers in Mobile. By the time the strikes entered their third month, the company had permanently replaced all 2,100 strikers.

Under the NLRA, the company is not required either to prove it needs the concessions to "maintain full production" or to submit the dispute to an arbitrator. Under the law the company can lock out the union workers and replace them with so-called "temporary replacements." I say "so-called" because there is no time limit in the definition of "temporary replacements." For example, in 1984 BASF Corporation in Louisiana locked-out 370 members of the Oil, Chemical and Atomic Workers Union for five and a-half years.

The Right to Strike?

The National Labor Relations Act maintains that workers have a right to strike, but the company has the right to permanently replace them the following day. Under that definition, the right to strike is often little more than the right to risk or sacrifice your job. In respect to labor law, what have workers gained since slavery was abolished?

---

* Over the years unions have had some success in getting clear language defining the responsibilities a worker assumes when she takes a particular job. These "work rules" are seen by management as impediments to their ability to manage.

**Twelve years later International Paper Co. closed the plant.

What is really going on here is that the NLRB makes sure the company has the ability to maintain production *as the company defines production*, the familiar conclusion being: what is good for International Paper Company is good for America. But what about workers. Are not workers' needs and rights part of the national interest? If the employers' "right" to maintain production by hiring permanent replacements is more important than the right of workers to keep their jobs, improve their standard of living and obtain justice in the workplace, than what other conclusion can one draw? Nothing, except to say "what is good for the corporation is good for America," and the people who work for the corporation are just second-class citizens. Real wages today are lower than they were in 1973; meanwhile the rich have been getting richer.[2] These losses are occurring as the power of unions has been decreasing.[3]

The cycle of poverty is a myth. The cycle of wealth is real. Wealth cycles from the working class, who create it, to the ruling class who use it to leverage more and more wealth from the working class.

Labor Board Cases Can Never Go To a Jury

The legal process the NLRB sets up is also a travesty of justice. Union workers do not get the opportunity to take a case to a jury of their peers. The process works like this: first, the NLRB Regional Director decides if a union or a worker has a case. If the Regional Director makes that determination and the company disagrees, then the case is put in front of an administrative law judge who makes a ruling. If either side disagrees with the outcome, the decision can be appealed to the full Board and then in some cases the Board's ruling can be appealed into the federal court system. It is impossible for a worker or union to get a labor board case in front of a jury of our peers. Throughout the whole process, lawyers and judges (the Board is made up of lawyers) decide the fate of workers. The notion that workers can routinely receive any justice under this system is ludicrous.

Article 3, Section 3 of the U.S. Constitution, "Trial of all crimes shall. . . be by jury."

Unit Size

Unions tend to organize workers in relatively small units in comparison to the overall number of workers employed by today's large corporations. There are a number of reasons for this, the most important being unions often lack the resources to organize a large group of workers, under the current NLRB rules. The result is that once organized, workers at one facility of a large corporation then have to stand alone to negotiate a contract.

"The Pool"

Even when a union gets contracts at a large number of a company's mills, as is the case at the International Paper Company, the individual mill contracts often expire at different times. To rectify the problem, in 1987 union workers at International Paper Company tried to increase their power by "pooling" the votes of individual bargaining units as their contracts expired. Thus some locals would work without a contract, but in the end they would create a company wide bargaining unit. In this way, over an extended period of time, they could respond to the company's contract proposals as a group, but the Board declared the "pool" process illegal.[4]

When the workers
sell their Labor
to the capitalists
or accumulators
 of Labor
they allow the
 capitalists
or accumulators
 of labor
to accumulate
 their labor.
And when the
 capitalists
or accumulators
 of Labor
have accumulated
 so much
of the workers'
 Labor
that they do
 no longer.
Find it profitable
to buy the
 workers' Labor
then the workers
can no longer sell
 their Labor
to the capitalists
or accumulators
 of Labor.
And when the workers
 can no longer sell
their Labor
to the capitalists
or accumulators
of labor
they can no
longer buy
the product of
their labor.
And this is what
the workers get
for selling
their labor.

Peter Maurin, The Catholic Worker

Union Certification

One of the best ways to illustrate the changes unions have undergone since the creation of the National Labor Relations Board is to look at the process by which the Board certifies a union as the collective bargaining agent for a unit of workers, a process known as the union certification election. But first, to get some perspective, let us review the process of "perfecting the organization" which unions relied on before the implementation of the National Labor Relations Act.

Perfecting the Organization

Before 1935, workers, exploited by their employer, would form an organization for their own protection and benefit either with or without the help of a national union. The workers' sources of power were their skills, unity and community support. Their potential for unity started among co-workers and extended into the community and other workplaces. For example, union printers, when they got strong enough, wouldn't use paper that wasn't union made. Union paper mill workers, when they were strong enough, wouldn't let their employers hire non-union construction workers to build new facilities. Workers would boycott stores that sold non-union made goods and union workers refused to handle struck goods. Workers set up and controlled apprentice and training programs as a source of pride and power. Unions were a major political force in some states, counties, cities and towns. They had real newspapers and schools. They put on plays, sponsored dances, ran co-op stores and credit unions. But most importantly, union members relied on each other to survive.

I was once assigned to help a Republican get re-elected to the Maine Legislature. I say "once" because she was the only Republican I ever helped. She had been endorsed by the Maine AFL-CIO because she always voted with labor on important issues. I asked her why she routinely went against her party and voted for labor. She said, "My father was a union steelworker and he was killed on the job when I was very young. My mother had no means of support, so the union took care of us and even put me through college. I will never forget the union until the day I die!"

Vote Union!

Today if you work in an unorganized facility and think you need a union, you call a union and ask for help. An organizer comes and explains to you the benefits of belonging to a union. You are told to get people to sign union authorization cards and when enough are signed, the NLRB will supervise an election to see if your co-workers want a union to represent them in bargaining with your employer. Through leaflets and house visits, you and the organizer try to convince other workers to sign union authorization cards. If enough are signed, the organizer goes to the company and makes a demand for recognition, which is routinely turned down by the employer. The organizer than goes to the NLRB and presents the authorization cards. If there are enough cards, the Board calls for a hearing to determine what the appropriate unit will be. That is, a hearing is held to decide who will be able to vote in a "union certification election."

Early on in the process the company almost always hires a union-busting law firm to advise it on how to keep the union out. Workers are then inundated with written propaganda and their supervisors' verbal barrage, aimed at scaring workers so they won't vote for the union. The union responds with leaflets telling workers how bad the company is and how much they need a union to represent them.

The problem this process presents for unions is not the fear that the company creates to get workers to vote against the union. The problem is the election process itself. In this process workers are not organized into a union. They are asked, under much duress, to vote for the union or against it, but they don't really become part of a union until they win the election. Even then, it is often questionable if they are really part of a union or are just paying dues to a collective bargaining agent. They are not organized around the idea that they will be part of a self-sustaining organization of workers that will be around whether or not the government requires the company to recognize their organization. They are organized around the idea that if a majority votes for a union, the company has to recognize that union, which in turn represents them. And, in a year, if the union you elected to represent you is not doing a good job, you can have another election and decertify the union. It is like a political election in which voters are asked to sit back and pass judgment on which party or politician they want to represent their interests in the legislature.

The whole idea of having an election to see if workers want to be represented seems surreal when you look at it this way. Suppose that in federal House of Representatives elections every two years, presidential elections every four years and senatorial elections every six years, people first voted on the question 'do you want to have a congress or a president?' Then if the majority wanted representation, a contest would ensue to see which candidate would hold the position. However, if a majority voted against representation, there would be no government, no representation and the only laws would be private corporate laws.

If 10, 20 or 1,000 workers form a union, the only elections held should be the ones where workers choose their union officers and when they vote to accept or reject a contract. The concept here is called "freedom of association."

THE COMPANY OWNS THE WALL!

But workers in this country don't have "freedom of association" or "freedom of assembly" or "freedom of speech" at the work site. So the organizing that takes place around an election is very superficial. It is a process in which the company holds most of the cards. After all, the company owns the property and signs your check!

The first union recognition election I participated in was in 1976 at a Converse Rubber Plant in North Berwick, Maine. Converse was a subsidiary of Eltra Corporation and close to 500 people were employed at the plant where I worked. During the campaign the company put up graphic anti-union posters all over the plant; whenever you looked up from your work, you had to see one. After a "captive audience" speech in which the mill manager said he wanted to see a fair election, I asked the personnel manager if the company really wants a fair election, why can't the union put up posters next to the company posters? He replied, "When the union owns the walls you can put up posters!"

The reason they're called "captive audience" meetings is they take place during working hours. So what choice do workers have but to go? Before the Supreme Court ruling in *NLRB v Virginia Electric and Power* (1941), captive audience meetings were illegal. In fact, the very act of an employer or their representatives even suggesting that they didn't think a union was a good idea was a clear violation of the Act.

These so-called free elections take place on private property where workers do not have freedom of speech or freedom of assembly. Under the NLRA, it is a protected activity for workers to talk union in non-production areas on non-production time. If you think that is free speech, you are living in corporate heaven. Union recognition elections are brutal. Workers are harassed and fired. Sure, you can go to the NLRB and ask it to take your case. But if the Board takes your case and you win, it may be years before you get your job back. Meanwhile, the company will have had the time to either get rid of, convert, promote or isolate the other organizers.

Now Get a Contract

What happens if you prevail against the tide and the union actually wins the election and the NLRB certifies the union as the collective bargaining agent? The next thing is to negotiate a contract. Here, once again the employers throw up a wall. According to Craig Becker, Associate General Council for the Service Employees International Union, "Five years after they had voted for union representation, over one-half of all workers were not covered by collective bargaining agreements. . . ." [5] Why? It's like training someone to participate in a political election and then sending them off to fight a war with some leaflets and no gun. The union still has not built an organization for the long haul, and the company never stops attacking. They continue to fire union activists and promote union sympathizers out of the bargaining unit. By the time the legal dust settles there is another election; only this time the union is decertified.

Is it any wonder that only 9 percent of the industrial work force in this country is represented by unions, or that the real wages of workers have been declining since 1973?

Vision and Power

Vision and power are two of the most important ingredients to the success of any movement. We cannot lay the blame for the lack of organized labor's vision on the National Labor Relations Board or the corporations. That is labor's responsibility, for it is only through a new vision that new power will spring. *It is labor's responsibility to let the sun shine through.*

The vision of the United Auto Workers in the 1930s was to organize all auto workers into one union with one set of demands. Like the Knights of Labor and the Industrial Workers of the World, unions that preceded it, the UAW wanted to break down the divisions among workers of craft, race, gender and skill. They wanted to eliminate competition among workers. And they wanted to eliminate competition among corporations to see which corporation could force workers to work longest for the least money.

The Pulp & Sulphite Union also organized on an industrial basis and to some extent across craft lines. The strike of 1908 taught them and the Paper Makers Union that they had to work together across craft, mill, and company lines to achieve a lasting victory for workers. That is what they accomplished in the strike of 1910.

Back then the Pulp & Sulphite Union held a wage conference every year to set the standards they would try to achieve across the industry. From 1910 until 1921, it was the standard they set at International Paper Company that the unions struggled to obtain industry-wide. The essence of what they fought for was to create an industry-wide bargaining unit that prevented paper companies from using wages as a competitive edge against other paper companies. That was their vision. Their power came from the leaders who articulated that vision and the members who lived it. They lived the vision by working together, struggling together, sharing their pain, their burdens and their accomplishments. They held parades and picnics, and wrote poems and songs. They gave support to the injured and weak. They mourned together, they organized together, they won and lost together. But the key was they got through it TOGETHER.

Remember the Oysters

Due to the success of unions in the 1930s, like the Longshoremen and Teamsters on the west coast and the Auto Workers and Mine Workers on the east coast, and to the passage of the NLRA, the paperworkers' unions were able to rebuild their membership base. This approach, in the short term, increased the number of members in the union. But it decreased the unions' power over the long term by destroying the knowledge of how to organize strong, meaningful, membership-based unions.

The problem with relying solely on others for your power and direction is that someday you end up like the oysters in the poem who relied on the carpenter and the walrus to take them for a walk. The oysters were not in control of their own destiny, so they ended up easy pickings. The paperworkers relied on the potential power of other unions as a source of their own power, and on the NLRB to give them direction and protection. As more and more unions went the way of the paperworkers, fewer and fewer unions retained their own power.

The National Labor Relations Board directs unions and workers into an endless legal maze. The Board will not teach workers about the historic struggles of their sisters and brothers. Nor can the Board challenge workers to build a democratic society based on justice and equity for all people.

Workers must do that!

Endnotes

1. Local 82 of the Amalgamated Clothing and Textile Workers Union — Shoe Division and Local 1915 of the International Brotherhood of Painters and Allied Trades.

2. According to U.S. Department of Labor, Bureau of Labor Statistics data, in1982 dollars, nonsupervisory workers earned about $40 a week less in 2002 than they did in 1973. "The top 20 percent of American households have more that 80 percent of the country's wealth" (Bradsher).

3. In 1979 unions in the U.S. had 21 million members, 24 percent of the workforce in 2002, 16.1 million, 13.2 percent of the workforce (U.S. Bureau of Labor Statistics).

4. Getman 206-207.

5. Becker 104.

MINE WORKERS AND PAPER WORKERS 12

IF THERE IS NO STRUGGLE THERE IS NO PROGRESS

by Frederick Douglass

If there is no struggle there is no progress. Those who prefer to favor freedom and yet deprecate agitation, are people who want crops without plowing up the ground, they want rain without thunder and lightning. They want the ocean without the awful roar of its many waters.

This struggle may be a moral one, or it may be a physical one, and it may be both moral and physical, but it must be a struggle. Power concedes nothing without a demand. It never did and it never will. Find out just what any people will quietly submit to and you have found out the exact measure of injustice and wrong which will be imposed on them, and these will continue till they are resisted with either words or blows, or both. The limits of tyrants are prescribed by the endurance of those whom they oppress.(From the West India Emancipation Speech, August 1857.)

MINE WORKERS

The United Mine Workers of America (UMWA) is a union whose traditions can be traced back before the Knights of Labor. The Mine Workers experienced many of the same problems that affected other national unions in the United States after World War II, including corrupt leadership, suppression of dissidents within the ranks, closing of work sites and a substantial loss in membership. Despite all their problems, the membership tossed out the corrupt leadership, regrouped and in the late 1980s was able to draw on their militant history to beat back a major corporation's attempt to gain the same kinds of concessions workers in the paper industry were unable to resist.

PITTSTON

In 1989, the year after the Paperworkers were defeated by the International Paper Company, the United Mine Workers of America took on the Pittston Coal Company and won. Unlike the president of the Paperworkers, the president of the Mine Workers is elected directly by the membership. Unlike the Paperworkers, the Mine Workers remember and embrace a history of struggle that goes back to the beginning of the labor movement. This enabled the Mine Workers to build a movement that united the membership and beat the Pittston Coal Company despite a national political climate and economic conditions similar to those under which the Paperworkers were beaten by the International Paper Company.

THE DIFFERENCE

The Mine Workers utilized sit-ins and sit-downs to galvanize the membership and the public in support of their struggle. They faced serious jail time and fines large enough to financially bankrupt the union many times over. But the union was not destroyed; it got stronger. The difference between the two unions was that the Mine Workers put together a plan that involved the entire union and put the resources of the entire union into the struggle. The leadership of the Mine Workers didn't think of their union as a bank account. They saw their union as an organization of workers, families and communities. Such an organization could survive even without a bank account. A Mine Worker official declared during the strike: "The union existed before the company and will exist after the company. They can take away our building and bank account and we will still have a union."[1] The Paperworker leadership didn't see it that way.

THE PAPERWORKERS' BOTTOM LINE

I once mentioned to an official of the Paperworkers that Local 14 was considering a sit-in at the International Paper Company mill gate in Jay. He made it clear the Paperworker leadership would oppose any activity that might subject the national union to lawsuits by the company for restraint of trade. Time and again during the strike, the national union leadership made it plain that the bottom line of the strike was the amount of money in the national union's treasury. They were blind to the concept that it isn't a bank account that holds a union together. The idea that the strike had the potential to build a movement powerful enough to stop the corporations' assault on workers' wages, hours and working conditions was beyond their comprehension. They never studied the way the paperworkers of 1910 used the strike to build a union, so the concept of "Perfecting the Organization" never entered their minds.

PASSIVE LEADERSHIP

In 1987/88 Wayne Glenn, the president of the Paperworkers, didn't go to a single meeting of a local union to urge members to join the pool, reject contracts and join the strike. He was the epitome of an absentee leader. He came to one mass meeting in Jay during the 16 months of the strike. These weekly meetings were the heart and soul of the struggle against concessions for workers, not just at International Paper but throughout the entire paper industry. In fact, several paper company executives publicly stated that they negotiated non-concessionary contracts because they didn't want anything like Jay to happen in *their* mill towns. Glenn was asked to come to mass meetings on more than one occasion, but it wasn't until he was scheduled to be at a staff meeting in Maine that he ran out of excuses and found the time to come.

Activist Leadership

In Jay, the local union brought in people from all over the state, the region, the country and the world to attend and speak at its mass meetings. The locals were solely responsible for creating the solidarity around the strike. In contrast, the leadership of the Mine Workers not only attended the mass meetings but helped organize them. The Mine Workers brought in union members, their families and their allies from all over the world to attend mass meetings and stay at a Solidarity encampment they built to bring people together for the duration of the strike.

The Mine Workers, from the bottom up, saw themselves in the historical tradition of struggle that was the very essence of the union. The Paperworkers from the top down, through its bureaucracy and out through the pores of its International Representatives, with very few exceptions, did not understand or acknowledge a union as anything but a collective bargaining agency.

Perfecting the Organization

If the Paperworkers had the historical understanding and commitment that the Mine Workers have, the Paperworkers could have won the strike of 1987 against the International Paper Company. And if they hadn't won that battle, they would have gone a long way in "perfecting the organization" and in so doing built a base from which a victory could have been won.

But, would that have been enough?

Endnote

1. Green interview.

COMING FULL CIRCLE: MILITANCY IS NOT ENOUGH

13

LABOR AND MANAGEMENT

Today people in the labor movement don't talk about Labor and Capital. We talk about labor and management or industrial labor relations. Management represents capital in the workplace and society. Labor has forfeited to Capital any claims over the direction of society and it has conceded to management all issues except those of wages, hours and working conditions.

And now that the unionized industrial work force is near 9 percent of all industrial workers (meaning, of course, that 91 percent of the industrial work force is unorganized), management prerogatives also dominate the arena of wages, hours and working conditions. For example, in my own experience beginning in the mid-1970s, unions proposed changes to management during contract negotiations. The question was how much of its agenda would labor get. But by the mid-1980s, the roles were reversed and it was the management agenda that was negotiated and the question of how many concessions labor would have to make dominated negotiations.

THE END OF THE DISCUSSION

Although the polarization between Capital and Labor largely ended in the early 1920s, a public discussion which questioned the role of Capital continued into the late 1940s. Then the cold war's withering hand, McCarthyism, clamped a suffocating lid on any serious questioning of the role of Capital being the dominant force in the United States. To argue otherwise was declared unpatriotic by the people in power. Many who continued to raise questions lost their jobs and some were put in jail. Unions that refused to buckle under were run out of the AFL and the CIO. "In 1949 the CIO purged unions representing 900,000 workers, for refusing to purge themselves of Communist leaders and support government policies such as the Marshall Plan."[1]

The big question being debated at the end of the nineteenth and beginning of the twentieth centuries was in whose interest would the society be run; who would set the parameters of the political debate; who would define education; would the culture be consumer oriented or cooperative? Capital or Labor? Labor in the paper industry eventually lost the war in the strike of 1921 and in 1937 accepted a very subservient role to Capital.

The victory of the Mine Workers over the Pittston Coal Company in 1989 in many ways recalls the CIO victories of 1937. Militant, innovative tactics, mass organizing and the use of a strike to build a union were all activities utilized by the Mine Workers, as they were used in the 1930s by a number of radical and militant unions and by many unions earlier, like the paperworkers in 1910. But can we build a new labor movement based solely on the tactics of the CIO? I think not.

Roosevelt and Lewis

The New Deal and the CIO greatly improved the lives of many people for many years. But what happened to the CIO?* What happened to the New Deal? Neither Franklin D. Roosevelt nor John L. Lewis wanted a revolution. Roosevelt wanted to create a full-production economy which he hoped would eliminate poverty and at the same time keep most of the profits going into the bank accounts of the people he grew up with, the already rich. Lewis wanted the laboring people he grew up with to get a bigger piece of the pie, but he didn't question in any fundamental way the role of the rich in running the society. In 1935 Lewis said:

> Labor does not want to undertake management's task. It does not want to run anybody's business.. .. .. It wants a place at the council table when decisions are made that affect the amount of food that the family of the worker may eat, the extent of the education of his children, the kind and amount of clothing they shall wear, the few pleasures they may enjoy.. .. ..[2]

Labor, by Lewis's standard, produces and consumes but does not contest the role of the corporate elite in deciding what is produced or consumed; what and how fast resources are extracted; what our children are taught; who our heroines and heroes are, and most importantly, in whose interest is the society run, Capital or Labor?

The Legacy of John L. Lewis

There are some problems inherent for labor in the John L. Lewis militant-but-still-pro-capitalist view of the world, but it is a view commonly held by

* Although the New Deal made things better for many people it also kept the relative status of people the same. For example, Social Security guarantees that low-wage workers will have lower retirement incomes then high wage workers and unemployment insurance payments are based on income before a layoff, thus preserving once again the relative status of people based on income, not need.

many labor people. Michael Moore, in a chapter of his book, *Downsize This!*, entitled "Why are Union Leaders So F#!@ing Stupid?," reveals this limited view of labor's role in the larger society:

> The list of Union Leadership Stupidity could fill the rest of this book. It is sad for me, on a personal level, even to have to write this chapter. My uncle Laverne participated in the Great Flint Sit Down Strike of 1936-7, which resulted in the first contract for the United Auto Workers and was, as historians have written, the beginning of the modern-day labor movement across America. Because of what my uncle and others fought for over the years, families like mine were able to live in homes that we owned, go to a doctor whenever we were sick, get our teeth fixed whenever they needed it, or go to college if we chose to — all thanks to the union.[3]

Moore heard the stories of 1937 from his uncle and could see how that struggle was responsible for putting food on the Moore table. He goes on to suggest that the problems for organized labor began when Reagan fired the air traffic controllers in 1981 and the unions didn't respond. Moore's view is based on living memory.

Michael Moore is a clever, creative, thinking person, yet the parameters of his thinking have been defined by the information presented to us by the corporate culture and organized labor's complicity in that culture. We are led to believe that the Knights of Labor and the I.W.W. were marginal, quirky movements, not part of the *modern* labor movement. The modern "mature" labor movement supposedly began with the National Labor Relations Act of 1935 and the CIO in 1937. Underlying this view of labor history is the assumption that the Knights and the I.W.W. were naive and immature because they dared to challenge the hegemony of the corporate state, while the AFL and the CIO were mature because they accepted the role of the corporate elite in directing our society and determining the future of our culture.* The leaders of the AFL and the CIO saw the role of the state as limited to decreasing the severity of corporate harm and providing a safety net for the working class.

We Want to Feel the Sunshine

The membership of the Knights and the I.W.W. saw material goods as necessities — not as the purpose of existence. They *rejected* the notion that most human activity should be geared toward creating consumer goods. They

* Hegemony is the predominant influence of one state over others. In our case it is the corporate state and culture over all other states and cultures.

rejected a society where those who *invest* money should earn a profit and those who *risk and invest* life and limb should only receive a wage. The Knights and I.W.W. fought for "8 hours for work, 8 for rest and 8 for what we will" while Carnegie, Rockefeller and Chisholm had people working 12-hours a day, six or seven days a week.

> We mean to make things over,
> We're tired of toil for naught,
> With bare enough to live on,
> And never an hour for thought;
> We want to feel the sunshine,
> And we want to smell the flowers,
> We're sure that God has willed it,
> And we mean to have eight hours.
> We're summoning our forces
> from shipyard, shop and mill;
> Eight hours for work, eight hours for rest,
> Eight hours for what we will!
>
> *from a poem by I.G. Blanchard written in 1866 which was put to music and used by the Eight-Hour League*

The radical unions rejected the notion that corporate managers like Chisholm, Carnegie, Ford, Du Pont, Morgan, Frick, Scott and Rockefeller should define for the common people the very essence of life. But the CIO accepted corporate hegemony and so modern corporate/labor history has us believe that the modern labor movement began with the CIO. Michael Moore, on the one hand, recognizes that without organized labor his family would not have been able to afford a modest home, send him to college or have his teeth fixed. But on the other hand, what about the future: what will we do with our lives and communities after we get our teeth fixed?

Does an expanding economy raise all boats?

On what stone tablets is it written that an ever-expanding economy, which extracts more and more natural resources and pollutes more and more of our Mother Earth's environment, raises all boats? In fact, only the boats of 20 per cent of our society are being raised and future generations will pay the price. Who still believes that an expanding economy run with higher and higher levels of technology will ever create enough "good jobs" so that everyone who wants one will have one? And who today believes that the less taxes the rich pay, the more they invest, the more jobs are created? Where is it written that freedom is the right to choose between 25 kinds of overpriced breakfast cereal grown by farmers who are almost always in debt while the corporate managers take the profits? Where is it inscribed that freedom is the right of General Motors Corporation to sell cars in Japan, as if the Japanese aren't capable of making their own cars? Who really benefits when cars made in this country are shipped to Japan and when cars made in Japan are shipped here? Whose world benefits when freedom and liberty become the right to exploit one nation's workers by forcing them to compete with another nation's workers? And whose freedom and liberty were trampled on when the managers of General Motors Corporation made the decision to scrap the public transportation system in this country, made up of trolleys and trains, and replace it with a private system made up of huge *public* highways and *private* cars?[4]

Is "Union Leadership Stupidity" the Problem?

The labor movement of the CIO never fundamentally questioned the liberal vision of Franklin Roosevelt, nor did the AFL-CIO of the 1980s ever raise any fundamental objections to the conservative vision of Ronald Reagan. Both eras provided excellent opportunities for organized labor to challenge the direction of the society advocated by those in power. But in neither case did labor throw down the gauntlet.

Michael Moore rightfully praises the CIO for winning material gains for many working class people in this country. But he shouldn't blame the problems of organized labor today on "Union Leadership Stupidity:"

> Our labor leaders today seem to have forgotten an important lesson we learned in elementary school: if you give the bully what the bully wants, the bully doesn't become your friend — he just wants some more because he knows he can get it! The only way you stopped the bully back in school was to stand up to him and face him down.[5]

The UAW-CIO faced down the GM bully in Moore's home town of Flint back in 1937. The UAW members took the property of the General Motors Corporation and held that property until the managers of the corporation recognized the union as the bargaining agent for the workers in that General Motors' plant — for six months.

But Moore understates the problem. The corporate elite are more than just bullies. The school yard bully is not the teacher, principal, superintendent of schools, local school board, state school board and the Federal Department of

Education and more all rolled into one. The managers of the General Motors Corporation may well be bullies and they may well direct corporations that make goods and provide services. But more importantly, **it is these managers who define what all the rest of us do with our lives.** These corporate managers control the direction of our society and the fate of the planet.

Militant Tradition: Circumscribed Goals

Moore draws on a militant tradition of action which was used to obtain important but highly circumscribed goals. The goals were to obtain a decent standard of living and respect on the job for union members. The program was to get company recognition and a good contract. The unions utilized the militant tactic of the sit-down strike to force the corporations to share more of the wealth with some of the workers who created it.

Meanwhile, the corporate managers were allowed to continue to consolidate more and more wealth and power and the unions were not. Workers got more money, better benefits and better working conditions (Moore got a decent roof over his head, teeth fixed, and a college education). The corporations had Taft-Hartley and the Prudent Man rule passed by Congress to prevent the unions from using the deferred wages of workers, pension and benefit funds, to buy the industries in which they work. Thus, the most important institution of the working class, the union, is prevented from consolidating economic wealth and power the way labor's chief adversary, the corporate elite, does.

Labor's New Sit-Down

In the 1970s, Ray Rogers presented organized labor with an updated version of the sit-down strike utilized by the CIO in the 1930s. Rogers' program is called the *Corporate Campaign* and its goal is to help unions prevail in disputes with management. Rogers makes it crystal clear that a campaign will be launched and will be constantly ratcheting up, eventually leading to the financial destruction of the corporation, unless it bargains in good faith with the union. Before the corporation is broke, he reasons, it will raise the white flag.

The strategy is to find the corporation's weakness and the union's strength and match the two of them up. Rogers maintains that the union's strength is its people and the corporation's weakness is the weakest link in the interlocking interests that support the corporate structure. He asks: Can the product be effectively *boycotted?* Can ties to the financial institutions that support the corporation be *severed* by organizing the economic power at the disposal of union members and their allies? Can moral, political and social power be utilized in ways to *isolate* the corporation from its suppliers or customers? Or as Rogers says, "We will organize a campaign to take the pain off the picket line and put it in the board room."

Rogers' strategy has two key components: one is identifying the corporation's weak links, and the other is maximizing the union's strength. Then the union and its allies apply their maximum strength to the corporation's Achilles tendon. But after Rogers' success in utilizing Corporate Campaign tactics against the notoriously anti-union textile corporation, J.P. Stevens Company in the late 1970s, organized labor proceeded to isolate him.

Some national unions started their own corporate campaign departments so that they could tell their members, "Don't go to Rogers, we can do it in-house." However, most of these departments were and are understaffed and poorly funded, and the campaigns they run often demoralize the very members they are supposed to be activating.

Organized labor's inability to implement the corporate campaign concept reflects the fact that most of today's union leaders have trouble dealing with the idea that unions can only survive if they challenge the existence of a corporation. The way they see it, it's the corporation that feeds the membership, not vice versa.

Boycott: Not My Company

The early stages of the 1987/88 paperworkers' strike provided a good example of this phenomenon. The local union leadership from the four mills involved pushed the national leadership of the United Paperworkers International Union to declare a boycott of International Paper Company products, but the national leaders declined.

Six months into the strike, under great pressure from the striking locals, the UPIU hired Ray Rogers to organize a corporate campaign against International Paper Company.* As part of the campaign, Rogers suggested not only boycotting International Paper Company products, but also products of companies like Anheuser Busch and Coca Cola that shared directors with the International Paper Company. The UPIU went along with the boycott of Anheuser Busch and Coca Cola products, but would not agree to boycott the International Paper Company itself. As some members of the UPIU executive board explained their reasoning, "We negotiate with IP at 22 other primary mills and many secondary mills. Workers at only four locations are

* Previously, in 1983 the UPIU had hired Ray Rogers to mobilize a "Corporate Campaign" against the International Paper Company. The campaign was called off when the CEO of International Paper Company agreed to terms acceptable to Wayne Glenn and his executive board.

out. We still have to negotiate with this company at all these other locations, so if we boycott the whole company it will sour our relations at the other mills." The reasoning was like the old adage that says, "If a hungry tiger asks you to give up your arm to satisfy his hunger, how long do you think it will be before the tiger comes back for the rest of you?"

Almost 14 months into the strike, in August 1988, the UPIU gathered in convention as it did once every four years. The convention directed the leadership to support the strikes against the International Paper Company in every way possible and a boycott of the corporation was finally announced. However, the UPIU leadership was simultaneously laying the ground work to end the strike.

Peace Accord

From the International Paper Company web page, history section:

"The strikes and lockout ended in the fall of 1988. In 1993, the company and the union signed a "peace accord" and agreed to return to the cordial relationship of earlier years."

After the strikes were over and all the workers had lost their jobs, the national union negotiated a pension agreement in 1993 with International Paper Company that had a side agreement in it called the "Peace Accord:"

> .. ..The management of the company, the international union, and its local unions, recognize that a return to the cordial and generally cooperative atmosphere that has characterized their relationship over the years is preferable to the adversarial climate of recent times.
>
> All parties recognize that the long-term economic competitiveness of the corporation is in the mutual interests of the employees and the shareowners; and the parties recognize that major ingredients of the economic success of the corporation are the skills, knowledge, and commitment of the employees.
>
> .. .. .. Specifically, the parties to this Memorandum, agree that attempting to resolve differences over safety and environmental issues by working together is preferable to calling in outside agencies or individuals.. .. .. The union further agrees that the corporate campaigns engendered in the past against International Paper will cease upon ratification of this agreement and will not be renewed so long as the company adheres to this agreement.

It is important to note that because of safety and environmental concerns raised by the unions during the strike, International Paper Company paid a $2.2 million dollar fine and was found guilty on five felony charges based on environmental violations cited by the federal Environmental Protection Agency. The state of Maine also fined the company heavily for environmental violations, and OSHA charges brought during the strike resulted in huge fines and required the corporation to invest over $50 million to update safety protection (not only in the struck locations but in many other mills).

The power the union had demonstrated with its safety and environmental offensive was abandoned to the hungry tiger when it signed the "Peace Accord." The UPIU leadership supported this agreement even though it could not make any inroads with the corporate managers in speeding up the return to work of any of the permanently replaced strikers or in gaining back any of the deep cuts workers took in wages and benefits during the period of the "adversarial climate."*

The UPIU surrendered its arms when it signed this "Peace Accord." Since then the company has continued to chop off little pieces whenever it chooses, meeting little or no resistance from a defeated and still discouraged union membership. The membership has little faith in its own ability to resist, much less in its national union's ability to negotiate anything but more concessions, more "Peace Accords."

President of PACE (formerly the UPIU) Boyd Young, who had represented some of the International Paper Company locals during the 1987/88 strike and became a UPIU Vice-President and chair of the International Paper Company Union Council in 1988 said in 2001: "We cannot come to terms with this industry, either nationally or around the world, unless we find a strategic way to deal with IP." The statement was made at an international meeting of labor leaders representing workers in the paper industry. PACE agreed to lead an international network of International Paper Company unions. "There is no question that if we win an equitable relationship with IP, we will be in a position to help improve conditions for paperworkers anywhere in the world," concluded Young.[6] Getting bigger will only help if the units you are bringing together are strong. Signing international Peace Accords won't help build a strong union anymore than signing national ones will. The changes necessary to build a union movement are not quantitative; they are qualitative.

* Nine years after the strike in Jay was called off, 200 of the 1,250 people who originally struck had not been rehired by International Paper Company. The significance of this number increases when you consider that many of the former strikers had taken a pension buyout from the company, given up recall rights, died or retired since the strike ended.

You Can't Go Back

It appears that what the leadership of the UPIU (now PACE) wants is a return to the "good old days" when the militant activities of other unions forced companies into the arms of the UPIU. Unfortunately for the UPIU, there are no militant unions out there organizing paper mill workers. Moreover, the UPIU leadership has no plans to build a power base out of its membership. The same can be said for most AFL-CIO unions. A strong active membership is a threat to them.

Another factor that makes it hard for union leaders to follow through with a Rogers Corporate Campaign strategy is that he sets the union on a path that will lead to the destruction of a corporation if it doesn't come to terms with the union's demands. The destruction of the company goes against the grain for most union leaders because the unions' bottom line in our society, as defined by the National Labor Relations Act, is to deliver industrial peace to the corporation, not to threaten or contest the existence of those corporations. And as a UPIU International Vice-President once said to me, "We can't destroy this company, we have contracts with it at other locations."

Mass Mobilization

The other problem union leaders have with Rogers' Corporate Campaign is his emphasis on the mass mobilization of the membership. Union leaders who are trained to sit down with management to negotiate contracts are for the most part very ill at ease with an active membership. In fact, many find an active membership threatening. There are several reasons for this relationship between the leadership and the membership, but the most important is the way most unions are structured.

Union Structure

Most local unions are structured democratically, and the membership directly elects the local leadership.* Most locals hold monthly meetings where decisions are made concerning everything from the allocation of the local's resources to the involvement of the local in political races. However, once you move past the local level, the institution is set up to suppress the potential involvement of the membership and to discourage or exclude the rank-and-file from real decision making. In his book, *The Paper Rebellion,* Harry Graham makes the argument that the lack of internal democracy was the rea-

* There are many exceptions to this statement, but it is true for most of the unions I had contact with during the 1970s and 1980s. However, in the 1990s there has been a trend to consolidate local unions into district and regional councils and boards that remove the membership from directly participating in the decision-making process.

son 20,000 members of the Paper Makers Union and the Pulp & Sulphite Union broke away in 1964 and formed the Association of Western Pulp and Paperworkers. Graham explains,

> A new union was formed which gave dissatisfied workers a chance to change their representation. This independent (non-AFL-CIO) organization was founded emphasizing internal debate, dissent, and criticism in contrast to the more general trade union practice which seems to equate dissent with disloyalty and criticism with treason.[7]

The organization most often identified when people think about unions is the AFL-CIO, an organization which speaks for unions. The AFL-CIO is an organization of unions, not a union, although its structure mirrors that of many unions. What follows is a description of the hierarchy of the Paperworkers Union (UPIU now Paper, Allied-Industrial, Chemical and Energy Workers International Union — PACE) and the AFL-CIO.*

The local union members directly elect local officers, including the local's president, through a secret ballot process in which the polls are open for any member to vote, similar to any federal, state or local government election. Delegates to official conventions are elected at local meetings. Once every four years, the UPIU holds a convention at which it elects its national and regional officers. The delegates to these conventions vote by weighted ballots. So a president of a 1,000-member local will often control a thousand votes, and the president of a local with 86 votes will have only 86 votes. The campaigns run by those seeking national office are conducted not amongst the membership at large but among those local leaders who cast large numbers of votes. The national union will then send delegates to the national AFL-CIO convention, which is conducted in much the same manner as the UPIU convention, only in this case it is even further removed from actual rank-and-file union members because the high-ranking officials of national unions cast almost all of the votes.

The UPIU pays dues, per capita, to the AFL-CIO on about 250,000 members and therefore the delegates to the AFL-CIO convention from the UPIU will carry 250,000 votes. If there are 12 million members in AFL-CIO-affiliated unions and some of those unions have more than 1 million members, 10 or so people will determine who is the president and what are the policies of the AFL-CIO. The result is that only a handful of people decide who speaks

---

* Every union is a little different in how it is set up. Although the UPIU is somewhat typical, it should be pointed out that a few unions like the Mine Workers, the United Electrical Workers and now the Teamsters have their Presidents elected directly by the membership.

for labor in this country. And that handful is generally pretty far removed from the everyday life and pressures of actual rank-and-file union members.

Structure Excludes Members

Putting aside the question of democracy and fairness in this process, there is another important consideration. Union members who are excluded from a process are less likely to identify with that union than they would with a union that included members in its process. An election process in which leaders are elected by the actual rank-and-file membership forces the leadership to speak and interact directly with the membership, thus creating the possibility of a strong and interdependent link between the membership and leadership of organized labor. When the leadership runs in an election in which the votes are weighted, the constituency of that leadership is the small group of people who cast a large number of votes.

Consequently, people seeking leadership in these union organizations go to the meetings, interact with, and learn the skills necessary to win elections among the union's elite, and therefore usually have no reason to learn the skills of mass public debate and organizing to keep their jobs. This contributes to the scarcity of high union officials who are able to speak effectively at public meetings, motivate and activate their own membership, or deal appropriately with the media. These leaders are spawned in an institution where they play an insiders' game. *That is what the institution of organized labor trains its leaders to be: inside players and inside politicians, not mass organizers or mass leaders.*

Militant — Radical — Democratic

Random House Dictionary, 1980

mil-i-tant.	1. Vigorous active and aggressive, esp. in support of a cause.
rad-i-cal.	1. Of or going to the roots or origins of a matter. 2. Favoring drastic political, economic, and social reform.
de-moc-ra-cy.	1. Government in which supreme power is vested in the people and exercised by them or their elected representatives.
	2. Political or social equality.

If we want to live in a democracy we must make a radical departure from the vision of the CIO and the AFL. That vision is based on the assumption that we can have a democratic society where some are very rich and powerful and everyone else — if they belong to a union — can gain economic security. But as the Knights of Labor said, you can't have a democracy when one group accumulates profits and the other works for wages. Because the one that gets the profits will use that money to become more and more powerful and that power will be exercised and accumulated in both the political and the economic realm. That is why union membership in this country peaked in the 1950s and real wages for 80 percent of the population have been declining since the early 1970s, along with the quality of family life and the environment. That is why our governments are being privatized and our public schools and public radio are now covered with corporate advertising. As the rich get more powerful they use that power to take more from the rest of society. To invade more of our space, to exploit more of our labor. With our hospitals, prisons and welfare programs privatized and corporatized, the rich are even profiting from our misery.

That is why today we need a new framework to view our history and a militant, radical, democratic movement to resist the tyranny of today's corporate elite. This movement needs to build a collective vision of a new society based on the needs, aspirations and survival of the inhabitants of this planet. This would lead us to a society very unlike the one we have today which is based on the *liberty* of the few to exploit the many.

Where do we begin? How do we start to build a militant, radical, democratic labor movement that will be part of a mass movement to end the tyranny of the corporate elite? A mass movement to build a new human society that will live in harmony with itself and the other species with which we are intertwined on this planet, and upon whom our very survival is dependent?

ENDNOTES

1. *Who Built America* Vol. II. 501.

2. "Labor and Cooperatives" 4.

3. Moore 129.

4. *Taken for a Ride,* a videotape by Klein and Olson documents the way the managers of General Motors Corporation systematically destroyed the public transportation system in this country so they could create a larger market for their cars by encouraging the middle class to abandon the cities for the suburbs, thus making our population totally dependent on automobiles, buses and trucks.

5. Moore 131.

6. *PACEsetter,* Vol. 3, No. 121. January/February 2001.

7. Graham xv.

Vision	A vivid mental image, the power to imagine.
Theory	An organized system of accepted knowledge that applies in a variety of circumstances to explain a specific set of circumstances.
Strategy	An elaborate and systematic plan of attack.
Tactic	A plan for attaining a particular goal.

14

SO WHAT IS TO BE DONE?

The Waves

Gains in union membership during the twentieth century came in two waves. The first began in 1900 when membership was just under a million and slowly increased until World War I when it took a leap. It reached 5 million in 1920 and then declined over the next 13 years finally bottoming out in 1932 at 3.4 million. The numbers picked up again dramatically beginning in 1936 at 3.6 million, going to 5.7 million by the end of 1937. The number of members continued to increase and peaked at 22.2 million in 1975. However, as a percentage of the workforce, unionization peaked in **1953 at 26 percent and by 2002 dropped to 13.2 percent (16.1 million members).**[1] Obviously, if organized labor doesn't do something soon to reverse this trend, it will go the way of the dinosaur.

When social scientists or mechanics are confronted with a problem they apply a theory to solve it. For example. You're driving down the road on the way to a union meeting and your car stops. If you knew nothing about cars, you wouldn't have a clue as to what the theory is that encompasses the automobile. So you would look at the tires and then open up the hood because you have seen other people do that. But that wouldn't do you much good.

However, if you were familiar with automotive theory you would know that a number of systems — like the electrical, fuel, and drive train — taken together make up an automobile. If you understand the theory behind how a car operates, you next need a strategy that addresses the current problem.

From similar experiences, coupled with your knowledge of automotive theory, you know it is likely that your car has a problem in the fuel or electrical systems. So your strategy is first to check the things that usually cause a car to stop running: lack of gas or a dead battery.

If that doesn't solve the problem you will then have to examine the two systems they initiate, namely the fuel and electrical systems, and follow them as they go to the car's cylinders to see why there isn't ignition at that point.

The next step is to devise a series of tactics that will test your strategy. Let's start with the electrical system. The first tactic is to turn the ignition key: if it turns the starter you know that the battery is ok. The starter turns. Next question. Does the car have gas? Look at the fuel gauge. The fuel gauge shows you have a full tank. Now the gauge could be wrong and you decide to see if it's working by moving up a step in your strategy. You pull out a spark plug to see if gas is getting into the cylinder. There is gas in the cylinder. So you move back to the electrical system and see if electricity is getting to the spark plug by having a companion turn the ignition key to see if a spark jumps across the plug gap. No spark. AHA! So you know the problem has to be between the spark plug and the battery.

Next you check the fuses between the battery and the spark plug: VOILA! A fuse is blown. So you replace the fuse and start the car.

As you drive down the road your first thought is relief that the car is running and you will be on time to the union meeting where you are scheduled to speak on tactics, strategy, theory and vision in today's labor movement. You now know that you can use the automotive experience to shed some light on the concepts of tactics, strategy and theory but what about vision?

Now historically we didn't just go from walking to riding in cars. Some one had to have the vision of the automobile in the first place, right? No, first we domesticated animals and then figured out the wheel thing and then someone had the *vision* to combine the two and came up with the horse-drawn cart. Many years later someone invented the internal combustion engine. This enabled the person who had the *vision* of driving down the road in a cart without a horse to combine the cart and the internal combustion engine to create the first horse-less wagon or automobile.

The Vision of Organized Labor

Organized labor presently operates with a vision. Labor's vision pictures a society where workers join unions. The unions in turn negotiate wages and benefits for them with their employers. The good wages workers earn allow them to live decent material lives and the union requires the employer to create a healthy workplace: no grief from the boss and no toxins in the air.

This vision **assumes** a society where corporate managers make the decisions about what is to be made and how it is to be made in the plant. And on the outside **it accepts a system** in which corporate financed think tanks, media, foundations, trade organizations (their unions) and political contributions are used to control the political process and determine how the overall society will function, not just the one in the workplace. This form of society is called "corporate capitalism," and this form of unionism is called "bread and butter" unionism.

Now out of "bread and butter" unionism comes a theory to get more bread from corporate capitalism. But in order to utilize this theory unions need members and so they must also confront the problem that unions can't add members at a rate faster than the losses due to run-away-shops and increased productivity.

Organized labor's activity aimed at dealing with the problems of adding *membership* and getting more *bread* is presently based on the belief that we live in a democracy governed by the rule of law and labor obeys the law. The theory is if you don't like the laws, organize and change them through the electoral process. This is the operating theory of the AFL and it came to be at the American Federation of Labor's 1947 convention when Mine Worker President, John L. Lewis urged the delegates to challenge the Taft-Hartley "Slave Labor Act" in the street, not just in Congress. Lewis thundered:

> This Act was passed to oppress labor, to make difficult its current enterprises for collective bargaining, to make more difficult the securing of new members for this labor movement, without which our movement will become so possessed of inertia that there is no action and no growth, and in a labor movement where there is no growth there is no security for its existence, because deterioration sets in and unions, like men, retrograde. . . . I wonder what built up the labor movement in this country? Was it protecting laws and statutes that protected the organizers of our movement when they went out to the meetings? Oh, no! The founders of our Federation had no such protection. They had to fight for the right to be heard. They had to fight for the right to hold a meeting, and men had to sacrifice and sometimes die for the right to join a union. . . . And what you are doing today [by complying with Taft-Hartley] — you are repealing the 1919 injunction policy . . . and you are humbling yourselves in abasement before the return of government by injunction.[2]

Despite the AFL's position that Taft-Hartley was unconstitutional, AFL Secretary-Treasurer George Meany claimed that it was "the law of the land" and must be obeyed. Thus the theory, as articulated by Lewis, that labor is bound to disobey un-constitutional laws was abandoned. The new theory has tied labors' body and soul to the mast of a sinking ship because the theory is based on a lie. Workers don't have rights. Therefore if workers are to organize they must at times break the law. Because it is the law — and the power of the state behind the law — that prevents workers from organizing. While Lewis said labor needs to keep fighting in the streets, Meany was content to live off a large membership that would soon begin a half-century decline. A decline labor has yet to reverse.

CURRENT STRATEGY

Labor's strategy to organize new members is centered on the NLRB and has two parts. The first goes like this; urge members to vote for which ever Democrat is running for President. When elected, the Democratic President will appoint good people to the Board and they in turn will make it easier for labor to organize.

Carter's 4 years	1.8 percent = -.45 per year
Reagan's 8 years	1.8 percent = -.23 per year
Bush's 4 years	1.0 percent = -.25 per year
Clinton's 8 years	2.3 percent = -.29 per year

Now that part of the strategy hasn't worked very well. Let's look at the presidential years from 1976 through 2000, of which 12 have been Republican and 12 have been Democratic. When Carter was elected in 1976 union membership as a percentage of the workforce stood at 20.4 percent. When he left office it was 18.6 percent, when Reagan left office it was 16.8 percent; with Bush it was 15.8 percent and with Clinton 13.5 percent. In fact, the loss has been *greater* under Democrats (4.1 percent) than Republicans (2.8 percent) during the last 24 years.

The other part of labor's strategy is to convince more and more unions, at all levels, to put more and more resources into bringing new members into the unions. It is amazing that as corporations have been consolidating power with the active support of the government and the courts, unions still continue to bring in new members every year. But unions aren't organizing fast enough to increase their total numbers or the union percentage of the workforce.

The good news is labor is very effective in the things it does day to day — its tactics. Unions do *organize* workers. Unions do *represent* workers. Unions do *motivate* members to vote in political elections. But these tactics are based on a false vision of America, followed by a strategy that has failed to halt labor's decline. Which means that current labor tactics amount to putting a finger in the hole of the proverbial dam.

So maybe it's time for union people to take another look at labor history and the visions of society that were painted and strategies that were utilized in the past. Visions and strategies coupled with tactics that inspired workers to build powerful unions.

Labor History

Since the coming of the Europeans to North America, labor's good times have been few and far between. In the colonial period 30 percent of the population lived in a condition of involuntary servitude, 20 percent were slaves and 10 percent were indentured servants.[3] Early colonial law placed a maximum wage on "free labor."[4] After the revolution, in 1806, shoemakers in Philadelphia formed a union for their own benefit and improvement. A Pennsylvania judge, in the Philadelphia Cordwainers Case ruled the formation of a union was a combination of workers engaged in a conspiracy.[5] The court proclaimed, "A combination of workmen to raise their wages may be considered in a twofold point of view: one is to benefit themselves .. .. .. the other is to injure those who do not join their society. The rule of law condemns both." The shoemakers were found guilty of conspiracy — so henceforth the *act* of forming a union was considered a conspiracy.

By 1842 the courts had eased up a bit on workers' organizations, saying in *Commonwealth of Massachusetts v Hunt*:

> Associations may be entered into, the object of which is to adopt measures that may have a tendency to impoverish another [another being the employer], that is, to diminish his gains and profits, yet so far from being criminal or unlawful, the object may be highly meritorious and public spirited. The legality of such an organization will therefore depend upon the *means* to be used for its accomplishments.[6] (italics added)

In other words, the court held that unions were not in and of themselves conspiracies. But if the members acted other than as individuals (for example, to raise wages), then it would be up to the court to decide if the union acted as a combination engaged in a criminal conspiracy.

Later in the nineteenth century, the boycott was declared illegal in many states. Between 1880 and 1931 the courts issued more than 1,800 injunctions against strikes. By 1920 the courts had struck down roughly 300 labor laws.[7]

Meanwhile the courts were extending "human rights" to labor's antagonist, the corporation. In 1819 the Court said corporations had constitutional protection under the contracts clause in *Dartmouth College*.[8] Then in 1886 the Court declared in *Santa Clara* that the corporation was a "person" entitled to the 14th Amendment rights of due process and equal protection (this after denying women the right to vote under the same Amendment). Adding insult to injury, the 14th Amendment, which was added to the Constitution to protect the rights of former slaves, was used mainly to protect and enlarge corporate privileges. Supreme Court Justice Hugo Black pointed out in 1938 that,

> Of the cases in this court in which the Fourteenth Amendment was applied during the first fifty years after its adoption, less than one-half of one percent invoked it in protection of the Negro race, and more than fifty percent asked that its benefits be extended to corporations"[9]

THE GOOD NEWS

The reason there was a maximum wage set on labor in the colonies was because workers organized to demand higher wages. Slaves and indentured servants were beaten and tortured by their owners to intimidate them into not trying to escape or organize rebellions.

Indentured servants from Europe and slaves from Africa provided the founding fathers of our nation the labor to exploit the natural resources of North America. This primarily involuntary labor brought wealth and power to the founders, men like George Washington, Thomas Jefferson, and James Madison, enabling them to *free themselves* from the British monarch. But the colonial slaves and servants also fought to be *free*, not from the British king, but from *their* masters. In response to unfree laborers' attempts at *freeing themselves* from their *masters* the Framers put the following in the Constitution of the United States:

> Article IV Sec 2. [3] No person held in Service or Labour in one State, under the laws thereof, escaping into another, shall, in Consequence of any regulation therein, be discharged from such Service or Labour, but shall be delivered up on Claim of the Party to whom such Service or Labour may be due.

If slaves and servants hadn't been escaping into freedom, the founders wouldn't have found it necessary to put the returning servants clause in the Constitution.

If labor hadn't been actively organizing politically to pass labor laws to protect and promote the interests of the working class, the courts wouldn't have had to find those laws unconstitutional.

If labor hadn't been successful in organizing against the corporate tools of the black list, yellow dog contract, company store, spies, gun thugs, internal security police and outright murder, the legislatures and the courts would not have had to intervene on behalf of the corporate class with their injunctions, fines, prison terms and sometimes — as with the Haymarket martyrs, Joe Hill, Nicola Sacco and Bartolomeo Vanzetti — the death sentence.

If labor hadn't been successful organizing strikes and picket lines, then the courts wouldn't have found it necessary to issue injunctions, enforced by local police, county sheriffs, state police, state militia and federal troops to break the strikes and bust picket lines.

If the boycott hadn't been an effective labor tool, then the legislatures wouldn't have found it necessary to outlaw it.

If strikes weren't effective, then the corporate managers wouldn't have invented scabs.

If the closed shop hadn't been important to the "house of labor," then Congress wouldn't have outlawed it.

The fact that organized labor survives is a testament to perseverance of the working class.

So what inspired "the common people" over the centuries to persevere, to hold on, to keep on inventing new avenues of solidarity and struggle?

Vision and Hope

Tactics, strategy and theory don't motivate people, but a vision of a better world, nation, community or workplace does.

After a quarter of a century in the labor movement I have observed that most people who vote for a union do so because they have *hope* that with a union the place they work will be a better place. They have a vision that with a union they will be treated with respect both on the job and in the paycheck. It's like politics. Most people get involved because they have hope that they can make some small or large part of the world a better place, a more just place, for themselves — or if not for them, at least for the next generation.

Slaves and servants wanted to be free to participate in determining their future. They had a vision and they fought to create it as they did in 1728 when a group of slaves broke free of their shackles and built an African village and established a government near the present site of Lexington, Virginia.[10]

The goals of the 1830s labor movement — the ten-hour day and public education — was based on a democratic vision. Labor people argued that if they were to build a democracy, they had to be educated, and to be educated they needed time to go to school. So they fought for the ten-hour day and free public education not as benefits in and of themselves but as conditions necessary to bring about a democratic form of government under which there would be no slavery (chattel or wage), no one would be hungry, and greed and competition would be replaced with cooperation and love. Worker solidarity would lead the way.

They had a vision.

When the Pulp & Sulphite Union fought to build their organization in the early part of the twentieth century John Burke said:

> .. .. .. every stand of the working class against the criminal greed and avarice of the capitalist class is a step forward in the inspiring march of the laboring masses towards their final complete emancipation.[11]

Building on the Positive

There have been a number of positive steps taken by labor in the twentieth century to project the vision of a broad based labor movement in the popular mind. Thanks to these steps, some of that vision has been, at least for a while, encapsulated into law. Let's look at two of them — industrial democracy and the "glorious labor amendment."[12]

Industrial Democracy

During World War I, government support of workers in war-related industries was in sync with government propaganda that promoted political democracy overseas by supporting industrial democracy at home. As a result, labor organizations involved in the war effort increased their membership and power.

The War Labor Board was created by President Wilson to prevent labor-management disputes from curtailing war-related production. This Board conducted workplace elections to decide not which unions — *but which workers* — would form a workplace council to represent the workforce. The situation is similar to a political election. We vote for an individual who is running for office. We don't vote to see if we will have a government or not, because the idea of government and representation is assumed. If we don't assume workers' representation then we are left with the reality that most working people will be employed under a condition of involuntary servitude.

The War Labor Board adjudicated disagreements between the worker councils and management through an arbitration procedure conducted by the Board. If a corporation failed to go along with a Board decision, President Wilson used his war time powers to nationalize the corporation. On the other hand, if workers failed to abide by a Board ruling, the President sent the troops.

Although the War Labor Board didn't promote existing unions per se, anymore than the election process promotes parties, the reality was that the people who ran for the worker representative positions were often union leaders and these people invariably won because they were *organized*. Other workers then joined the union because it was the union, in fact, that was negotiating with the company for them. Hence, as union power increased, so did union membership.

This new-found power and the expectation of what could be achieved after the war was expressed at the 1919 and 1920 conventions of the United Mine Workers. They adopted resolutions which included:

- nationalization of the coal mines;
- public ownership of the railroads;
- legislation to make interference with union organizing a criminal offence;
- the thirty-hour work week;
- national health insurance;
- release of all political prisoners;
- the demilitarization of the United States;
- the formation of a labor party.[13]

The Glorious Labor Amendment

After the passage of the 13th Amendment in 1865, many workers believed that this Amendment applied to them. They called it "the glorious labor amendment."[14] This Amendment states in Section 1: "Neither slavery nor involuntary servitude. . . . shall exist within the United States," and in Section 2: "Congress shall have the power to enforce this article by appropriate legislation."

The "labor amendment" became the charter of the labor movement. Many workers believed that to be free meant the condition of involuntary servitude was abolished both on and off the job, and it was the responsibility of Congress to make the Amendment a reality. They argued that to be denied freedom of speech, assembly and organization on the job, and not to be free to strike and boycott, was tantamount to involuntary servitude.

To those who said workers were free to quit their jobs they replied that these conditions existed on almost every job a laborer in the industrial age might obtain, and therefore the right to quit had little meaning as long as a person needed a job to survive. Samuel Gompers said, "The only workers not laboring under terms and conditions arbitrarily imposed upon them from a source wholly foreign to themselves, are the organized workers."[15]

As far back as 1908 it was the official policy of the American Federation of Labor that a union confronted with an unconstitutional injunction had an "imperative duty" to "refuse obedience and to take whatever consequences may ensue."[16] [17] But the courts, with few exceptions, took the corporate side against the "labor amendment" position and routinely enjoined strikes, picket lines and boycotts. Thus it was the courts, labor held, that violated the 13th Amendment.

After many years of labor defying unconstitutional injunctions, Congress lived up to part of its obligation under the 13th Amendment in 1932 passing the Norris-LaGuardia Anti-Injunction Act. This Act outlawed the "yellow dog"contract and stated in Section 1 that "No court of the United States . . . shall have jurisdiction to issue restraining or temporary or permanent injunctions in a case involving or growing out of a labor dispute, except in a strict conformity with the provisions of this Act." In Section 2 of Norris-LaGuardia, Congress described the existing relationship between labor and the owners of property in the United States:

> Whereas under prevailing economic conditions, developed with the aid of governmental authority for owners of property to organize in the corporate and other forms of ownership association, the individual unorganized worker is commonly helpless to exercise actual liberty of contract and to protect his freedom of labor, and thereby to obtain acceptable terms and conditions of employment . . .

Congress then went on to lay out the solution:

> . . . it is necessary that he have full freedom of association, self-organization, and designation of representatives of his own choosing, to negotiate the terms and conditions of his employment, and that he shall be free from the interference, restraint, or coercion of employers of labor, or their agents, in the designation of such representatives or in self-organization or in other concerted activities for the purpose of collective bargaining or other mutual aid or protection; therefore, the following definitions of and limitations upon the jurisdiction and authority of the courts of the United States are enacted.

In one legislative act the injunction was lifted off labor's back. The yellow dog contract was abolished and it was hoped that workers could now enjoy full freedom of association and self-organization "for the purpose of collective bargaining or other mutual aid or protection."

Labor now had its Magna Charta. *Freedom of association was the law.* The American Federation of Labor's goals of ending the labor injunction and having a law recognizing the right of workers to freely associate was on the books. Labor could organize, labor could boycott, labor could strike and employers could no longer rely on the federal courts for injunctions or for federal troops to interfere on the employer's behalf.*

Labor and the New Deal

Despite this legislative victory, not much changed. It was 1932, the country was in the Great Depression, labor was weak and the employers still held most of the cards because Norris-LaGuardia restricted government activity but didn't prevent corporations from firing union activists or forming company unions. President Franklin Roosevelt brought the New Deal to the country the following year. This New Deal included the establishment of a Labor Board.

The purpose of Roosevelt's Labor Board (and shortly thereafter the National Labor Relations Board) was to cut down on strikes, which hurt production, by providing a mechanism for union recognition and "good faith" bargaining. Unions were also encouraged because the Roosevelt Administration believed that as more workers were organized, wages would go up. Workers would then have more money to spend on goods and therefore more goods

---

* Norris-LaGuardia does not prevent state courts from issuing injunctions in labor disputes. However by 1941, 24 states had anti-injunction laws — Millis 647.

would be produced and the country would "grow" its way out of the Depression — the opposite of trickle down, trickle up.

However, just because the President supported unions and Congress passed the National Labor Relations Act, it didn't mean that large employers like the General Motors Corporation or the U.S. Steel Corporation were going to automatically recognize unions. At least not until the Supreme Court said they had to.

During World War I, President Wilson relied on his war powers to insure workers' rights. If a corporation refused to recognize a workers council the President had the authority to nationalize the corporation. The Court and Congress were out of the picture. But that was not the case during peacetime and so, come the 1930s, when the tactic of mass picketing failed to achieve employer recognition, workers on a number of occasions organized general strikes affecting entire cities and sit-downs.

As the depression continued to deepen, workers continued to organize. Their power increased to the point that President Roosevelt was beholden to labor for his re-election to the presidency in 1936. Labor had become so strong politically that when workers took over the General Motors Corporation factory in Flint, Michigan neither the governor of the state nor President Roosevelt sent troops to remove the strikers from the property of the corporation.

The great Flint, sit-down which began on December 30, 1936, ended 44 days later with the United Auto Workers union gaining recognition at 17 General Motors Corporation plants. After Flint the sit-down movement escalated and by year's end, 477 sit-downs had taken place affecting 400,000 workers.[18]

However, the Supreme Court ruled many of Roosevelt's programs unconstitutional until he threatened to stack the Court by instituting a retirement age and increasing the number of judges to a maximum of fifteen. The threat to "stack' the Supreme Court and escalating worker militancy, expressed through an expanding sit-down movement and political mobilization, forced the Supreme Court judges to rethink their long standing resistance to labor law. As a result the Supreme Court reversed itself and declared the National Labor Relations Act of 1935 constitutional in *NLRB v Jones & Laughlin Steel Corp* (1937). The court's opinion reflected the language of the Norris-LaGuardia Anti-Injunction Act:

> Employees have as clear a right to organize and select their representatives for lawful purposes as the respondent has to organize its

> business and select its own officers and agents. Discrimination and coercion to prevent the free exercise of the right of employees to self-organization and representation is a proper subject for condemnation by competent legislative authority. Long ago we stated the reason for labor organizations. We said that they were organized out of the necessities of the situation; that a single employee was helpless in dealing with an employer; that he was dependent ordinarily on his daily wage for the maintenance of himself and family; that, if the employer refused to pay him the wages that he thought fair, he was nevertheless unable to leave the employ and resist arbitrary and unfair treatment; that union was essential to give laborers opportunity to deal on an equality with their employer.

Furthermore, the Norris-LaGuardia Anti-Injunction Act passed the constitutional muster of the US Supreme Court in February 1938. Many believed this meant that workers were free to organize and strike; boycotts were legal, and permanent replacements were not. They thought it meant that the yellow dog contract and the dreaded labor injunction were history.

Then in May 1938 the Court took a big step backward, returning to its historic anti-labor, pro-employer self, in *NLRB v Mackay Radio*, ruling that permanent replacement of strikers by employers was legal: this turned the right to strike into the right to quit.[19] The Court went on in *NLRB v Virginia Electric & Power (1941)* to grant free speech rights to employers in union certification elections.[20]

Despite these legal setbacks the labor movement continued to grow by utilizing the NLRB's certification process until 1953. But the cumulative effects of employer opposition, Supreme Court rulings and the passage by Congress of the anti-union Taft-Hartley Act of 1947, over a presidential veto, combined to begin a slow decline in the percentage of the workforce represented by unions. This rate of decline increased in the 1980s when major corporate employers which previously had seemed to accept labor as a partner in the industrial labor relations process pulled the plug. This activity was brought into clear focus in 1981 when President Reagan fired the PATCO (air traffic controllers) workers.

In response to the firings of the air traffic controllers, the AFL-CIO organized a massive march and rally in Washington D.C. in which over 500,000 union members and supporters participated. The AFL-CIO called the day of the march Solidarity Day. I was one of many who came to Washington for Solidarity Day thinking that over the ensuing months the march would be followed up by real acts of solidarity with the PATCO workers. However, the leadership of the AFL-CIO didn't have a long-term strategy to get the PATCO workers their jobs back and very little organizational support followed the rally. This lack of solidarity on the part of labor helped to usher in a wave of corporate demands for labor concessions in times of record profits that followed the PATCO loss. Maybe it is time for labor to look again at the historic concept of worker solidarity, because in so doing labor might begin to reclaim its true meaning and role in building working class power.

SOLIDARITY

It is not uncommon at union rallies for union representative after union representative to stand up and say something like, "the one-thousand members I represent stand shoulder to shoulder, *in solidarity*, with the sisters and brothers on strike here today." The union rep then presents a check to the striking workers and everyone cheers.

That's great. But is it real solidarity?

No, it is *support.* This is because these same workers who are giving the money to their striking brothers and sisters will likely not refuse to handle the struck work (goods and services produced by scabs) of the striking workers, and the building trades workers will likely continue to honor the contracts they have with the company that instigated the strike or lock-out. That is not worker solidarity it is support. Such support is important because workers involved in solidarity activity need all the help they can get.

Solidarity means that workers must be willing to put themselves in a similar situation to those on strike. They would have to strike in solidarity by refusing to cross the picket line or by participating in a secondary boycott by refusing to handle struck goods. Real acts of solidarity are rare today, but there are exceptions.

One occurred during the Jay strike of 1987/88 when building trades members refused to cross the paperworkers' picket lines and in so doing lost years of work at the International Paper Company in Jay. Actions of support are common but real solidarity is rare. Why?

It is not because union workers are not up for the fight. It is because, in the short term, the threat of an injunction and the lawsuits that follow weigh heavily on the leadership of organized labor. But what should the leadership of organized labor do? How about putting the "glorious labor amendment" back on the front burner. The question is how?

First, labor must return to the pre-Taft-Hartley policy of defying unconstitutional laws. Second, labor must politicize the labor movement by trying to make the Democratic Party the party of labor. An all out attempt should be made to replace the party's corporate agenda with a labor agenda and retire the corporate identified party managers with leaders who will articulate a working class program. If labor is defeated in its attempt to make the Democratic Party a party of labor, than it can take the organization built trying to do that and form a Labor Party of its own.

Labor Party

In most other industrial countries, labor parties have influenced the political process over most of the twentieth century: the United States is an exception. In Spain and France, by law, workers have 30 paid vacation days. In Canada, workers have universal health insurance.[21] These conditions are not negotiated with management, work-site by work-site, but are rights defined by the legislative process. They are the law of the land.

In the United States, unions go one on one with employers to negotiate private contracts that cover wages, hours, benefits and working conditions. In other words, where there are strong labor parties, issues relating to fundamental human dignity are defined in a public legislative process where everyone who is human gets the benefits. This stands in sharp contrast to a private corporate process — as in the US — where only those who have the power at a particular plant, working for a particular company, get good benefits and wages. Furthermore, the workers at these facilities can never feel secure that they will keep the benefits they have worked for because every time their contract expires so might their wages, benefits and working conditions.

Health care is a human right and the US Constitution should reflect it! Worker representation should be embedded in every corporate charter issued by "we the people" through our state governments. It is time for labor to argue that worker representation is guaranteed in and protected by the 13th Amendment to the Constitution which ended slavery and involuntary servitude; by the 1st Amendment which is supposed to guarantee freedom of speech, assembly and association; and by the 14th Amendment which was passed to insure that all Americans — not just the wealthy — have due process and equal protection under the law.

The 13th Amendment outlawed the conditions of slavery and involuntary servitude and required that Congress pass laws to implement the ending of both. The promoters of the 1st Amendment argued that the purpose of free speech is to insure that people can participate in the process of governing. The 14th Amendment guarantees due process and equal protection to all people who come under the 13th Amendment.

It is time for labor to insist that Congress pass laws setting up a workers' council system at every workplace and conduct free elections to see who will sit on those councils. The idea that workers should vote for having a union should be relegated to the history book of bad ideas. And the very idea that the government should conduct elections for who represents people at work should be as common place as the idea that we have elections for who represents us in Congress.

Given labor's experience with the conduct of so-called "free elections" by the NLRB, it is time for organized labor to begin the struggle for a real system of democracy at work. A system built on the 13th , 1st and 14th Amendments. If these Amendments are to become real for working people, labor has to assume a process of governance that guarantees worker representation. A process similar to the one that existed for two short years in this country under the World War I Labor Board.

In order for working people to begin to make these demands, they have to regain the ability to think and act collectively.

The Notion of Collective Rights "An Injury to One is an Injury to All"

Most union people who have processed a grievance can relate to the following. A member wants a grievance filed because he or she didn't get overtime as laid out by the seniority provision in the contract. The shop steward explains to the member that once the grievance is filed it becomes the property of the union. The member says, "No way, this is *my* grievance. It's not the union that is being hurt, it is me. This is personal!"

To which the steward replies: "'An injury to one is an injury to all.' This is not about you, this is about us. The contract the company has isn't with you. It is with all of us as a body, as a collective. The foreperson may have messed with *you,* but the company violated its contract with *us.*"

Just as a grievance is collective so is striking. Striking isn't about quitting together. Rather, it is about being free to participate in a collective bargaining process. If you can be fired for participating in a collective process, *then you are not free.*

The right to boycott; to strike; to form political parties; to pass real labor laws; to file grievances and negotiate contracts — these are *collective rights.* They are created and maintained by concerted collective action.

For working people, collective activity is our past and the key to our future. It is our CONSTITUTIONAL RIGHT. As working people, it as our responsibility to exercise our rights!

So how do we get there?

Movement Forming, Recruiting, Educating and Politicizing

Lawrence Goodwin, in his book *The Populist Moment,* lays out four stages of democratic movement building:

- Create an institution where new interpretations that run counter to those of the existing institutions can be formed.
- Utilize tactics that attract masses of people.
- Educate new members.
- Demonstrate independent political power, what Goodwin calls "the movement politicized."[22]

Today within organized labor, "new interpretations," like those in this book, are materializing that run counter to those of the prevailing corporate state that union leaders now accept. Labor has in hand the knowledge and historical experience to "attract masses of people" and to educate them. As to "demonstrating political power," labor people have for over 170 years experimented with the creation of a labor party through which new ideas can percolate up, be expressed and put into actions that will define a society built on a truly democratic foundation.

To help facilitate Goodwin's four points there are several periods in the paperworker experience that bear looking at. The first comes out of the strike of 1910. Here a union used a strike and activist national leadership to build a collectivity of working people strong enough to exercise power on behalf of their constituents both in the community and the industry. To do this they had to put aside divisions among themselves, have an activist national leadership and include the community in their organization. During the strike, the paperworkers attracted masses of people and educated new members. After the strike they exercised independent political power.

Which brings us to politics. Working people have to understand the difference between playing politics and political action.

Playing Politics or Political Action

Playing politics focuses on the politician. It involves advising workers on whom to vote for, based on who will hopefully help (or not likely hurt) labor. Political action involves educating the working class about labor history and corporate culture. It involves organizing people to exercise their rights collectively to change the conditions they live under and the society they live in. Political action focuses on educating and organizing working people whereas playing politics focuses on influencing politicians.

I was at a meeting of local union leaders during the 1984 presidential primary campaign. A national AFL-CIO official said that labor was going to support Mondale. When asked why he responded that, "Mondale was with us in the past and besides he is going to win."

At the time I was a shop steward in the Painters Union. When I returned to work I gathered the 20 or so painters on the job and told them about the meeting and about the directive given by the AFL-CIO. My brothers were not exactly inspired and one asked, "Will Mondale be any better as President than Carter was, and if not, why bother to vote?" The AFL-CIO's expectation was that with Mondale they would have someone that would give them a few political appointments and not try to break labor the way the incumbent Reagan was. The AFL-CIO's program wasn't based on a vision which inspired workers, or for that matter which really included them. Therefore it failed.

Four years later, working people in Jay, Maine faced an environmental problem created by the International Paper Company. The Company was in violation of numerous environmental laws that affected the health and safety of mill workers and the community. The leadership of the local unions had to decide on what course of action to take. They could lobby state and federal politicians and hope for the best or they could develop a plan of political action and take it to the strike community that met, over a thousand strong, once a week.

A plan to get the town to pass a precedent-setting local environmental ordinance which would give the town of 5,000 people the same authority as the federal Environmental Protection Agency (EPA)

- to issue licenses for land, water and air pollution,
- to monitor and enforce the law, and
- to fine the company for violations,[23]

was taken to the strike community at a mass meeting where it was overwhelmingly adopted. After the plan was adopted by the mass meeting, educational materials were developed explaining why the ordinance was necessary and how it would work. Then a petition was circulated to put the ordinance on the town meeting agenda. Public hearings were held and a majority approved the ordinance at the following town meeting.

The Town of Jay then hired a person full time to administer the ordinance. Thereafter when an environmental problem came up townspeople didn't have to call the state capital or Washington to try to get environmental laws enforced, they called the town hall.

As a result of this political action organized by the union people of Jay, not only did they have a local course of their own to follow when the company violated environmental law, but both the state and the federal government started to do a better job. One result was the International Paper Company was found guilty on five felony counts and fined 2.2 million dollars for past environmental violations in federal court. But most important of all the people of the town, with union leadership, didn't play politics. They took political action.

Matt Wuerker

Today, labor people are confronted with problems greater than those the paperworkers faced in Jay back in 1988. But if labor combines the community organizations which the paper mill unions built in 1910 and 1987 with the national political organization and thinking of the CIO, and add to it the worldwide struggle for democracy, labor will find itself on the cutting edge of a movement. Once again labor people will be educating, organizing and leading the struggle to create the democracy of "We the People." Labor will make real the promise the Framers put into the Preamble of the Constitution, to "establish Justice" and "promote the general Welfare."

It is time for labor to bring to the fore the notion of collective rights by acting in real solidarity. It's time for labor to read the working class into the Constitution and drive the corporations out.

Endnotes

1. All data on union membership are taken from publications of the U.S. Department of Labor's Bureau of Labor Statistics; a National Bureau of Economic Research 1965 report by Leo Troy; and the Bureau of National Affairs (BNA). From 1930 to 1980 the data were taken from surveys sent to unions. After 1980, the method of collecting and interpreting the data changed. Also, the total number and percentage of organized workers vary according to who is included or left out. For example, for awhile workers over 14 were included and then the age for those included was increased.

2. American Federation of Labor, Proceedings 487, 490, 492.

3. Smith 336.

4. Morris 20.

5. Commonwealth (PA) v Pullis (1806).

6. Commonwealth (MA) v Hunt — 45 Mass. 111 (1842).

7. Forbath 38 and Fink 150.

8. Dartmouth v. Woodward — 17 U.S. 518 (1819).

9. Connecticut General Co. v Johnson — 303 U.S. 90 (1938).

10. *Who Built America* Vol. 1. 69.

11. Voelker 2.

12. Forbath 137.

13. Lichtenstein. Article by David Montgomery, Industrial Democracy or Democracy in Industry?: The Theory and Practice of the Labor Movement, 1870-1925 36.

14. Forbath 137.

15. Gompers 469, 472-73.

16. *American Federation of Labor* 254 "And we therefore declare that we will exercise all the rights and privileges guaranteed us by the Constitution and laws of our country, and insist it is our duty to defend ourselves at all hazards, and we declare that such be our action, taking whatever results may come."

17. Gompers 469, 472-73.

18. Fine 302, 331.

19. National Labor Relations Board v Mackay Radio — 304 U.S. 333 (1938).

20. Employer free speech case — National Labor Relations Board v Virginia Electric & Power Co. 314 U.S. 469 (1941).

21. Goodwin xviii.

22. Robinson 50.

23. Kellman. *New Solutions* 66-75.

CONCLUSION

Joe Hill

by Alfred Hayes

I dreamed I saw Joe Hill last night alive as you and me.
Says I, "But Joe, you're ten years dead."
"I never died," says he. "I never died," says he.

"In Salt Lake, Joe, by God," says I, him standing by my bed,
"They framed you on a murder charge." Says Joe, "But I ain't dead."

"The copper bosses killed you, Joe. They shot you Joe," says I.
"Takes more than guns to kill a man," says Joe, "I didn't die."

And standing there as big as life and smiling with his eyes,
Joe says, "What they forgot to kill went on to organize."

I dreamed I saw Joe Hill last night alive as you and me.
Says I, "But Joe, you're ten years dead."
"I never died," says he. "I never died," says he.

Joe Hill was an I.W.W. song writer, longshoreman, miner and lumberjack. While supporting copper strikers in Utah he was framed by the cops, found guilty by the courts and shot by a firing squad in 1915.

In the introduction, I put forward the proposition that in exploring the history of the paperworkers and organized labor, this book would answer three critical questions:

- What forces and events created today's institution of organized labor?

- What is it about the institution of organized labor that keeps labor from building a working class movement?
- What are the positive lessons from the past that workers can draw on to propel labor into the future?

On June 13, 1903 a strike began in the 20 paper mills of Holyoke, Massachusetts. Sixty-seven days later the strike of 4,500 people was lost and the Paper Makers Union that called it was down to 40 local members in good standing. The Vice President of the national union wrote the following to the people that had been involved:

> Vain are written decrees of wisdom, idle the sighing regrets of misled men and women. "Alas! It might have been," is an ejaculation that has escaped the lips of many a sad and disappointed member of the human family in after moments, and when his memory lingers along the rough path of the years and days faded and vanishing into oblivion of the past.
>
> Were we to allow ourselves to see nothing but disappointments staring at us in our recollections of the past, then indeed would the demon of despondency embrace us in his unyielding clasp, and the word hope, whose beautiful meaning has been the life-giving elixir of striving humanity through long ages past, lose its strong uplifting powers.
>
> We can and do see other things gainful to us because of our little portion of inherent wisdom which teaches us to light our future path with the flaming torch of bitter experiences, whose severe lessons have been indelibly burned into our minds and hearts.
>
> J. J. McCarthy[1]

Workers in the paper industry didn't let "the demon of despondency embrace" them in "his unyielding clasp." They took "the word hope" and "went on to organize." They came back from that defeat. At times they fought to make the 13th Amendment to the Constitution a reality as, for example when their contracts specified that the mills could run on Sundays only with the approval of the union. Later they fought for "industrial democracy" when in 1918 and 1919 they struck to enforce a decision of the War Labor Board. And in 1921 they struck the "American Plan" of the International Paper Company for five years. They lost the strike but kept their dignity and put the International Paper Company on notice that they would be back.

But for the last 66 years of the century the dominant unions in the paper industry have relied primarily on other unions for their survival. Initially it was the general strike on the West Coast followed by the threat of CIO unions organizing papermakers that encouraged the paper corporations to once again recognize the Paper Makers and the Pulp & Sulphite unions. This threat created by workers in other unions was primarily responsible for the huge membership increases of the unions in the paper industry.

Later as their membership numbers declined, the Paper Makers and the Pulp & Sulphite unions merged. Now they rely almost entirely on mergers to keep their dues paying membership numbers at a point where they can maintain a national office and staff. This trend is not unique to the unions that formed among paper workers: it is true for most of organized labor.

But, a new century has dawned finding labor in a world of hurt. *It is time* for labor to once again cast off the "demon of despondency," and reach back to claim "our little portion of inherent wisdom which teaches us to light our future path with the flaming torch of bitter experiences, whose severe lessons have been indelibly burned into our minds."

I know for some members of the Paper, Allied-Industrial, Chemical and Energy International Union (PACE) and for most pro-labor people, the information in this book may not have been easy to accept or digest. The history of the paperworkers, as it is for most workers, has been hidden from you for good reason.

For one thing, people tend to avoid looking at their defeats. For another, labor history is inspiring, and there were those in the union who over the years did not want an inspired membership — indeed, they preferred a docile, obedient and unquestioning membership.

But this history is part and parcel of the history of organized labor. Like the paperworkers who were defeated and almost destroyed in 1908 and came back together for their greatest victory in 1910, organized workers who have suffered the defeats that have occurred since the passage of Taft-Hartley can come back as never before — if they are prepared to celebrate the victories and look with clear eyes at the defeats.

Endnote

1. Burns 101.

APPENDIX

PAPERWORKERS' CHRONOLOGY

1884	The first union of paper makers, an association of highly skilled paper machine tenders called the Eagle Lodge, formed in Holyoke, Massachusetts.
1893	The American Federation of Labor (AFL) granted the Eagle Lodge and its sister organizations a charter. The new union took the name of United Brotherhood of Paper Makers (UBPM).
1898	Machine tenders form the International Paper Machine Tenders Union (IPMTU).
1901	Semi-skilled paper mill workers join the Federal Laborers Protective Union Local No. 9,555 in Chisholm (Jay).
June 16, 1902	International Paper Company signs a contract with the Federal Laborers Protective Union, the forerunner of the Pulp & Sulphite Union.
1906	Pulp workers hold national convention and form the International Brotherhood of Pulp, Sulphite and Paper Mill Workers (IBP-SPMW).
	The union has 3,700 members in 16 locals.
August 2/3, 1908	Paper Makers Union strikes the International Paper Company. Defeated, the union ends the strike on November 7, 1908.
1909	The AFL recognizes the International Brotherhood of Pulp Sulphite and Paper Mill Workers (Pulp & Sulphite).
March 6, 1910	Paper Makers and Pulp & Sulphite Unions strike the International Paper Company and end it victoriously on May 2.
1914	John Burke leads the New Hampshire Socialist Party as their candidate for Governor.
1917	John Burke elected President of the Pulp & Sulphite Union — a position he holds until 1965.
	The union had 10,000 members in 75 locals.
October 21, 1917	International Paper Company grants a 10 percent bonus.

July 24 -29, 1918	Strike over company's refusal to grant 10 percent increase ordered by the War Labor Board.
May 12–26, 1919	Strike over wages ended by order of the War Labor Board.
April, 14, 1920	One-year contract signed between the paper-making unions and the International Paper Company. The next contract was signed in 1937.
May 1, 1921	Paper workers across the U.S. reject companies' offer of 30 percent wage reduction, nine-hour day and elimination of overtime pay and 25,000 workers strike.
June 28, 1921	Unions vote to accept the Carlisle Group's arbitration proposal and return to work. International Paper Company does not agree to arbitration and the strike there continues for five years.
March 20, 1926	Unions defeated, call off strike which began in 1921 against the International Paper Company.
1933	Pulp & Sulphite Union has 5,684 members.
1934	In August, the paper industry on the West Coast recognizes the Paper Makers Union (PMU) and the Pulp & Sulphite Union.
1935	Pulp & Sulphite Union has 15,700 members, PMU 9,300.
1937	International Paper Company recognizes Paper Makers and Pulp & Sulphite Unions in its Northern, Southern and Canadian Divisions. Northern Division contract signed on October 28.
1938	Pulp & Sulphite Union has 37,900 members, PMU 23,000.
1948	Pulp & Sulphite Union has 120,400 members, PMU 56,500.
1949	The CIO purged unions representing 900,000 workers for refusing to purge themselves of Communist leaders.
1957	United Paper Workers of America and the International Brotherhood of Paper Makers (CIO unions in the paper industry) merged with the PMU — total membership 118,300. Pulp & Sulphite 162,200.
1958	Pulp & Sulphite Union has 165,400 members. PMU 121,600.
	United Wallpaper Craftsmen and Workers of North America (10,000 members) merge with the Pulp & Sulphite.
1959	American Wire Weavers Protective Association (10,000 members) merge with the Paper Makers.
1964	20,000 members of the Paper Makers and Pulp & Sulphite Unions quit them and form the Association of Western Pulp and Paper Workers.

1965	International Paper Company builds new mill in Jay, Maine called the Androscoggin Mill.
1967	Pulp & Sulphite Union has 171,000 members in 723 locals.
1972	Union workers at Androscoggin Mill vote not to be part of the Northern Division bargaining unit. (Date not certified by author although several local sources believe it was 1972.)
1972	Paper Makers and Pulp & Sulphite Unions merge to form the United Paperworkers International Union. Membership 310,000 U.S. membership 254,000 (Troy).
1978	International Paper Company sells the Otis Mill in Jay, Maine.
1980	UPIU membership 275,000. U.S. membership 217,000 (Troy).
1986	Paper Companies meet and agree to go after concessions from the unions in contract negotiations across the country. International Paper Company, Boise Cascade Company and the Georgia Pacific Company lead the charge.
October 29, 1986	International Paper Company President, Paul O'Neil, in a speech to the Maine Chamber of Commerce states that the company will be going after double time on Sunday in up-coming negotiations with union. They also want a lower tax rate, a decrease in the Workers Compensation rate and less environmental regulation.
December 16, 1986	Local 14 offers IP a one year contract extension. IP rejects the offer.
February 1987	Union notifies IP they want to start contract negotiations. IP responds, wait until April.
March 20, 1987	UPIU tells International Paper Company it will utilize Coordinated Bargaining tactics against the company for refusing to back off concessionary demands.
March 21, 1987	IP locks out 1,400 workers at its Mobile, Alabama mill.
April 21, 1987	Contract negotiations begin. Union offers two-year extension. IP rejects.
May 4, 1987	First mass public meeting of the strike held in Jay. Larry Funk, chair of the Mobile, Alabama negotiating committee, addresses the meeting.
May 31, 1987	Contracts between IP and the two unions expire.
June 4, 1987	The unions reject IP's contract offer and authorize the first strike at the Jay mill in more than 66 years. Ninety-eight percent of the membership vote, 97 percent vote to strike.
June 9, 1987	The unions reject IP's "final offer."

June 16, 1987	1,250 members of the two unions walk off the job and the strike begins. Strikers rally at State House — largest in 40 years.
June 24, 1987	First of 60 weekly mass meetings held during the strike usually attended by over 1,000 people.
June 24, 1987	Moss Point, Mississippi mill rejects contract proposal.
June 29, 1987	Scabs brought into Jay mill.
July 2, 1987	First food bank distribution.
July 3, 1987	Veterans against IP hold first vigil.
July 7, 1987	Superior Court Justice signs an order limiting the number of union pickets at the mill to 12 per gate and barring union members from assembling on a road where an earlier confrontation had occurred.
July 10, 1987	235 scabs in mill.
July 17, 1987	470 scabs in mill.
July 19, 1987	UPIU announces plan to increase strike benefits from $55 a week to $200. (Plan never fully implemented.)
July 20, 1987	506 scabs in mill.
July 23, 1987	Selectmen nix $4.5 million bond for IP.
July 28, 1987	UPIU fails to serve notice that they want to negotiate a new contract at the Hamilton, Ohio mill. Company also failed to serve notice so the contract was automatically extended for a year beginning September 28, 1987.
July 30, 1987	UPIU fails to serve notice that they want to negotiate a new contract at the Corinth, New York mill. Company does serve notice.
July 31, 1987	611 scabs in the mill. Over half the work force is permanently replaced. IP sues union for $750,000.
August 1, 1987	10,000 people rally and march in Jay to support strikers.
August 3, 1987	Rally held in front of IP corporate headquarters in Memphis, Tennessee.
August 11, 1987	Environmental ordinances are adopted at a special Jay town meeting. IP files a lawsuit challenging the ordinances.
August 12, 1987	UPIU President Wayne Glenn attends a weekly mass meeting. Glenn hails Local 14 in Jay as "a labor hero" and urges the rank and file to persevere.
	Strikers are eligible for unemployment benefits.
August 22, 1987	200 Jay workers attend rally in Lock Haven, Pennsylvania.

August 24, 1987	Delegations from Mobile, DePere, Lockhaven, Jay and the UPIU meet with local leaders and members in Pine Bluff, Arkansas.
September 6, 1987	Thousands attend Labor Day rally in Waterville, Maine.
September 10, 1987	IP offers double time at its Menasha, Wisconsin mill.
September 23, 1987	Pine Bluff, Arkansas locals vote to accept contract; major setback for striking locals.
September 24, 1987	Maine Governor holds talks with union and IP.
September 28, 1987	IP mill in Gardiner, Oregon votes to strike.
	Hamilton, Ohio mill gets one-year contract extension.
September 29, 1987	IP and union negotiators meet for nine minutes, adjourning without making progress toward settling the dispute.
	IP states all but 100 workers have been permanently replaced.
September 30, 1987	Menasha, Wisconsin mill votes to strike.
October 12, 1987	Democratic primary presidential candidate Jesse Jackson headlines a union rally in Jay for more than 3,500 strikers and their supporters.
	At the mill, IP implements the contract rejected by the unions.
October 27, 1987	OSHA fines IP $242,000.
October 30, 1987	Corinth, New York mill rejects contract offer.
November 21, 1987	More than 6,000 strikers and supporters rally and march to plant gate. U.S. Rep. Richard Gephardt, then campaigning for the Democratic presidential nomination, addresses the crowd.
November 23, 1987	Unions fined $10,000; held in contempt for July 7 injunction limiting pickets at the mill.
November 24, 1987	The Paperworkers' Union announces it has hired labor leader Ray Rogers to lead a "corporate campaign."
December 8, 1987	Corinth mill accepts contract offer.
December 16, 1987	Department of Environmental Protection fines IP $55,000 for August waste water spill.
December 17, 1987	Strikers launch a public appeal and announce plans to lead a caravan through New England to gain sympathy for their cause.
January 28, 1988	Eight people are sent to the hospital following a hydrogen sulfide leak at IP.

February 5, 1988	Nearly 4,000 residents of Jay and surrounding communities are forced to leave because of a 112,000-gallon chlorine dioxide leak at IP.
February 14, 1988	Seven are injured in a chlorine leak at the mill.
February 17, 1988	Gov. John R. McKernan meets with IP Chairman John Georges in Portland to discuss safety at the mill in the aftermath of the leaks.
March 9, 1988	A group of top state officials who inspected the mill following the February leaks announce that they found no problems serious enough to warrant ordering the plant shut down.
March 28, 1988	IP and the unions open national talks to establish a framework for settling labor disputes at the Jay mill and IP mills in three other states. Corporate Campaign put on hold.
April 17, 1988	Strikers reject IP's latest proposal, throwing copies onto the floor of a meeting hall in Jay. The key provision would have IP Co. hire strikers back someplace in the country within a year; the union insisted that the strikers get their jobs back.
April 30, 1988	More than 6,000 people march to the mill's gates to show support of the strike.
May 20, 1988	A replacement worker is killed when he is struck by fragments from a grinding wheel. The Occupational Safety and Health Administration begins an investigation.
May 21, 1988	Voters in Jay enact the Jay Environmental Control and Improvement Ordinance. The precedent-setting ordinance gives the town the power to issue licences, monitor and enforce the same environmental laws covering air, water and land pollution that the Federal Environmental Protection Agency does.
June 6, 1988	Ticonderoga, New York workers reject contract offer.
June 21, 1988	Jay voters authorize a $100,000 special appropriation to defend a new environmental ordinance.
July 28, 1988	OSHA cites IP for hundreds of safety violations at the mill. IP agrees to pay more than $870,000 in fines.
August 5, 1988	OSHA proposes $11,000 in fines stemming from the worker's death in May.
August11, 1988	Attorney General James E. Tierney files suit against IP in Superior Court, seeking fines for violations of "virtually every environmental law."
August 12, 1988	UPIU national convention held in Las Vegas votes to have the union do everything in its power to win the strikes against IP.

August 23, 1988	Paperworkers charge that the strike has cost IP stockholders $284 million in losses.
September 16, 1988	Roving picket hits the Ticonderoga mill, workers stay out for half a day.
September 26, 1988	Jay selectmen approve a property tax reassessment for the mill for the first time in 12 years, pushing the mill's taxes up by about 15 percent, to $5.5 million a year.
October 10, 1988	UPIU abruptly announces an end to strikes at IP mills in Jay, Wisconsin and Pennsylvania.
1988	Strikers hired back — 12.
1989	Strikers hired back — 115.
August 2, 1989	IP fined $990,000 for violation of Clean Air Act and will install $5 million in pollution control equipment.
1990	Strikers hired back — 104.
April 13, 1991	IP fined $885,000 by State for past environmental violations.
July 3, 1991	IP found guilty on five Federal felony charges based on environmental violations fined $2.2 million, second largest in U.S. history. Three weeks later President George Bush appoints IP CEO John Georges to the President's Commission on Environmental Quality.
1991	Strikers hired back — 70.
	Independent Workers of North America (cement workers) affiliate with UPIU bringing in 5,000 members in 74 Locals.
1991	UPIU membership 202,000.
July 1992	Local 14 is decertified by a vote of 361 to 616.
1992	Strikers hired back — 72.
1993	Strikers hired back — 39.
1994	Strikers hired back — 21.
	Allied Industrial workers merge with UPIU bringing in 45,000 members.
August 1995	Local 14 loses certification vote 373 to 571.
1995	Strikers hired back — 25.
1995	UPIU membership 233,000.
November 1996	Local 14 loses certification vote 392 to 511.

1996	Strikers hired back — 26.
1997	Strikers hired back — 40.
	UPIU membership 226,000.
March 30, 1998	533 strikers recalled to date, 173 to go.
June 4, 1998	Local 14 loses certification vote 316 to 576.
June 8, 1998	Referendum held in the Town of Jay on the environmental ordinance. Voters support keeping the ordinance by a margin of 2 to 1.
January 5, 1999	Oil, Chemical and Atomic Workers Union and the UPIU merged to form Paper, Allied-Industrial, Chemical and Energy International Union (PACE), membership 330,000.
June 6, 2000	The last lawsuit coming out of the strike of 1987/88 was settled when the company agreed to pay 20 union members, who were not included in the company's recall list, $2,000 each.

Note: Sources vary on union membership. The purpose of including membership numbers in this chronology is to show trends only.

Works Cited

Agee, Lynn and Glenn, Wayne. Meeting. UPIU Headquarters, Nashville, TN, October 10, 1988. (Author was present when Agee, General Counsel of UPIU, in the presence of UPIU President Glenn explained the President's constitutional obligation in calling off a strike.)

Altgeld, John P. "The Condition of Labor in This Country." Watertown, NY: *The Journal*, the official organ of the United Brotherhood of Paper Makers, Vol.1, No. 3, February 1, 1902.

American Federation of Labor: History, Encyclopedia Reference Book. Prepared and published by authority of the 1917 and 1919 conventions. Washington, DC. 1919.

American Federation of Labor, Proceedings of the Sixty-Seventh Annual Convention, 1947.

Annual Report of the Bureau of Industrial and Labor Relations, State of Maine. 1887, 1895, 1899, 1904, 1906, 1908, 1909, 1918.

Becker, Craig. Elections Without Democracy. *New Labor Forum.* Fall/Winter 1998.

Bowman, Scott R. *The Modern Corporation and American Political Thought: Law, Power and Ideology.* PA State University Press, 1996.

Boyer, Richard O. and Morais, Herbert M. *Labor's Untold Story.* NY: Cameron Associates, 1955.

Bradsher, Keith. "Gap in Wealth in U.S. Called Widest in West." *New York Times.* April 17, 1995.

Burke, John. Letter to Harvard University. November 25, 1949, on file at Littauer Library, Harvard University, Cambridge, MA.

Burke, John, editor. *Part of the Historical Record of the Labor Relations Between the International Paper Company and the International Brotherhood of Pulp, Sulphite and Paper Mill Workers.* Fort Edward, New York: Astoria Press, 1950.

Burns, Matthew J. *History of the IBPM.* Unpublished. Circa 1922.

Carbonella, August. *The Making and Unmaking of a Local Working Class in Jay/Livermore Falls, Maine 1900-1988.* Ph.D. Dissertation. Graduate School, City University of New York, 1998.

Chaplin, Ralph. *Bars and Shadows.* Ridgewood, NJ: Nellie Seeds Nearing, 1922.

Collier, William Miller. *The Trusts: What Can We Do With Them? What Can They Do for Us?* NY: The Baker and Taylor Company, 1900.

Dubofsky and Van Tine, *John L. Lewis.* NY Times Books Co., 1977.

Fine, Sidney. *Sit-down: The General Motors Strike of 1936-37.* Ann Arbor: University of Michigan Press, 1991.

Fink, Leon. "Labor, Liberty and the Law: Trade Unionism and the Problem of the American Constitutional Order," *In Search of the Working Class: Essays in American Labor History and Political Culture.* Urbana: University of Illinois Press, 1994.

Fitch, John A. "To L.W. Hatch, Chief Statistician, Bureau of Labor Statistics, Department of Labor, Albany, N.Y. "Preliminary Investigation into the Causes of the Strike of the Employees of the International Paper Company and the Events Connected Therewith." Albany, 1910. Ithaca, NY: Cornell University Labor Archives.

Forbath, William E. *Law and the Shaping of the American Labor Movement.* Cambridge, MA: Harvard University Press, 1991.

Getman, Jack. *The Betrayal of Local 14.* Ithaca, NY: ILR Press, 1998.

Gompers, Samuel. "Union Labor and the Enlightened Employer," *American Federationist.* Vol. 28. 1921, 469, 472.

Gonya, Wilfred. 1971 interview. Tape 600 Maine State Federation of Labor Collection, Northeast Archives of Folklore and Oral History, University of Maine, Orono.

Goodwyn, Lawrence. *The Populist Moment.* NY: Oxford University Press, 1978.

Graham, Harry E. *The Paper Rebellion.* Iowa City: University of Iowa Press, 1970.

Green, James. Interview with author. 1995.

Green, William. Statement on Executive Order 6580 (Feburary 3, 1934), AFL Records reel 57.

Gross, James. "Unionism in the Pulp and Paper Industry." *Labor History,* Vol. 5, No. 2. Spring 1964. Tamement Institute, NY, NY.

(A) Group of Men. Paper on file at the library of the United Paperworkers International Union. Nashville,TN.

(A) Guide to basic law and procedures under the National Labor Relations Act. Prepared in the Office of the General Counsel, NLRB. Washington, DC: U.S. Government Printing Office, 1962 Revised.

Gutman, Herbert G. *Power and Culture — Essays on the American Working Class.* NY: Pantheon Books, 1987.

(The) History Factory. *Generations of Pride 1898-1998. A Centennial History of International Paper.* N.p.: n.p., 1998.

International Paper Company After Fifty Years. International Paper Company, 1948.

International Paper Company Notice, October 18, 1917. In possession of author.

Kellman Papers. Special Collections Department, University of Maine Folger Library, Orono, Maine.

Kellman, Peter. "An Organizers Diary — The Environmental Ordinance Case in Jay, Maine." *New Solutions, A Journal of Environmental & Occupational Health Policy,* Vol. 4, No. 2, (Winter) 1994.

— "Jay, Maine Fights for Jobs & the Environment." *Labor Research Review,* Vol. 22, 1994.

Klein, Jim. Olson, Martha. *Taken for a Ride.* Film. New Day Films. 1996.

Knights of Labor Constitution & Declaration of Principles. 1878.

"Labor and Cooperatives: A Common History." *Community Economics.* Fall 1968.

Lewiston Sun [ME].

Lichtenstein, Nelson ed. *Industrial Democracy in America.* NY: Cambridge University Press, 1996.

Lewiston Evening Journal [ME].

Livermore Falls Advertiser [ME].

Lustig, Jeffrey R. *Corporate Liberalism.* University of CA Press, 1982.

Lynd, Staughton. "Truth In Organizing." *IMPACT.* Vol. 5, No. 11, February 1998.

— *SOLIDARITY UNIONISM: Rebuilding the Labor Movement from Below.* Chicago: Charles H. Kerr Publishing Company, 1992.

— *We Are All Leaders.* University of Illinois Press, 1996.

Millis, Harry and Montgomery, Royal. *Organized Labor.* NY: McGraw Hill, 1945.

Mitchell, John. *Organized Labor — Its Problems, Purposes and Ideals and the Present and Future of American Wage Earners.* Philadelphia: Bible House, 1903.

Moore, Michael. *Downsize This!* NY: Crown Publishers, 1996.

Morris, Richard B. *Government and Labor in Early America.* NY: Harper, 1965.

National War Labor Board, Docket #35, Employees v. Manufactures of Newsprint Paper. Report of Section to the Full Board. July 26, 1918, 6.

New York Times. [NY, NY]

Norris, John Esq. *The Paper Trust.* Boston, MA: New England Free Trade League, 1901.

PACEsetter. Publication of the Paper, Allied-Industrial, Chemical and Energy Workers International Union. Nashville, TN.

Palm Beach Post [Florida].

Pain On Their Faces. Ed. Peter Kellman. NY: The Apex Press, 1998.

Paper Makers' Journal. [Albany, NY].

Portland Press Herald [ME].

Poulin, Mickey. Interview with author 1989.

Pulp Makers President, *"Letter to B.H. Trethewey,"* 27 April 1906. Ithaca, NY: Cornell Labor Archives.

Quin, Mike. *The Big Strike.* Olema, CA.: Olema Publishing Company, 1949.

Robinson, Joe. "A New Campaign To Give Americans More Free Time Takes Off." *Utne Reader.* September-October 2000, 50.

Rumford Falls Times [ME].

Scontras, Charles A. *Organized Labor in Maine: Twentieth Century Origins.* Orono, Maine: The Bureau of Labor Education of the Continuing Education Division University of Maine, 1985.

Smith, Abbot E. *Colonists in Bondage: White Servitude and Convict Labor in America 1607-1776.* Baltimore: Genealogical Publishing, 1998.

Sturtevant, Reginald H. *A History of Livermore.* Lewiston, ME: Twin City Printery, 1970.

(The) United Paperworkers and International Paper: Ninety Years of Struggle. United Paperworkers International Union. Nashville, TN: 1987.

Troy, Leo. *Trade Unions Membership, 1897-1963.* NY: National Bureau of Economic Research, 1965.

U.S. Department of Commerce, Consumer Price Index.

U.S. Department of Labor, Bureau of Labor Statistics. *Current Population Survey.* May 1977-80; Annual Averages, 1983-1997.

United Mine Workers of America. *John L. Lewis — The Story from 1917-1952.* United Mine Workers of America, 1952.

Voelker, Keith Emery. *The History of the International Brotherhood of Pulp, Sulphite, and Paper Mill Workers, from 1906 to 1929: A Case Study Of Industrial Unionism Before The Great Depression.* Dissertation (1969). University of WI.

Who Built America?, Vol. II. NY: Pantheon Books, 1992.

Zieger, Robert H. *Rebuilding the Pulp and Paper Workers Union, 1933-1941.* Knoxville: University of Tennessee Press, 1984.

Zinn, Howard. "Eugene V. Debs and the Idea of Socialism," *The Progressive*, January 1999.

LAWS CITED

Adamson Act of 1916 (8 hr day railroad workers) — c.436, 39 Stat. 721

Clayton Anti-trust Act of 1914 — c. 323, 38 Stat. 730

Espionage Act of 1917 — c. 30, 40 Stat. 217, Title 1

Federal Child Labor Law of 1916 — c. 432, 39 Stat. 675

Interstate Commerce Act of 1887 — c.104, 24 Stat. 379

Maine Law 1887 Chapter 139. p. 121-125. Hours of Labor for Women and Children.

----,1891 Chapter 19, p.13. Labor Day.

----,1891 Chapter 303, p. 267. Conspiracy.

National Labor Relations Act of 1935 — c.372, 49 Stat. 449

Seamen's Act of 1915 — c.153, 38 Stat. 1164

Sedition Act of 1918 — c.75, 40 Stat. 553

Sherman Anti-trust Act of 1890 — c.647, 26 Stat. 209

Taft-Hartley Act of 1947 — c.120, 61 Stat. 136

POEMS AND SONGS CITED

Blanchard, I.G. "Eight Hours for What We Will." *Boston Daily Voice.* August 7, 1886.

Carroll, Lewis. *Through the Looking-Glass.* NY & London: Harper & Brothers, 1902, 68-73.

Hayes, Alfred and Robinson, Earl. "Joe Hill." 1934.

Hill, Joe. "The Preacher and the Slave." *The Little Red Song Book.* I.W.W. Ypsilanti, MI, 1995. (Appeared first in the 1912 edition.)

Maurin, Peter. "When Conservatives Try to Conserve." *The Catholic Worker.* NY. May 1998. Vol. LXV, No. 3, 6.

—, "When the Workers Sell Their Labor." *The Catholic Worker.* May 1997, 8.

RECOMMENDED REFERENCES FOR THE HISTORY OF UNIONS IN PAPER MILLS

Brotslaw, Irving. *Trade Unionism in the Pulp and Paper Industry.* Ph. D. Dissertation, University of Wisconsin, 1964.

Burke, John, ed. *Part of the Historical Record of the Labor Relations Between the International Paper Company and the International Brotherhood of Pulp, Sulphite and Paper Mill Workers.* Fort Edward, NY: Astoria Press, 1950.

Burns, Matthew J. *History of the IBPM.* Unpublished. Circa 1922.

Carbonella, August. *The Making and Unmaking of a Local Working Class in Jay/Livermore Falls, Maine 1900-1988.* Ph.D. Dissertation. Graduate School, City University of New York. 1998.

Cernek, Stephen, R. *Beyond the Return to Normalcy: The Decline of Organized Paperworkers, 1921 — 1926.* Dissertation — Muncie, Indiana: Ball State University, May 1978.

Graham, Harry E. *The Paper Rebellion.* Iowa City: University of Iowa Press, 1970.

Gross, James. "Unionism in the Pulp and Paper Industry." *Labor History,* Vol. 5, No. 2, Spring 1964. Tamement Institute, NY, NY.

(A) Group of Men. Paper on file at the library of Paper, Allied-Industrial, Chemical and Energy International Union. Nashville, TN.

Pain On Their Faces. Ed. Peter Kellman. NY: The Apex Press, 1998.

Voelker, Keith E. *The History of the International Brotherhood of Pulp, Sulphite, and Paper Mill Workers, from 1906 to 1929: A Case Study of Industrial Unionism Before the Great Depression.* Dissertation (1969). University of Wisconsin.

Zieger, Robert H. *Rebuilding the Pulp and Paper Workers Union, 1933-1941.* Knoxville: University of Tennessee Press, 1984.

INDEX

1877, Great Railroad Strike, 52*n*
1894 Pullman strike, 69, 72
1903 strike, paper workers, 168
1906 strike, paper bag makers, 35
1907 strike, Great Northern Lockout, 41–42, 46, 49, 50
1908 strike, paper workers
  cartoons, 43, 45
  chronology, 42–47, 170
  contract violations, 49
  mistakes made during, 61
  scabs, treatment after, 49
  union membership statistics before/after, 29
1910 strike, paper workers
  1987/88 strike compared, 63
  cartoons, 53, 58
  causes of, 50–51
  in chronology, 170
  culture of solidarity during, 63
  democratic movement building illustrated in, 161
  ending
    corporate strategy for, 59–60
    outcomes, 63–64, 66
    settlement, 60–61
    strike breaking strategies, 51
  expansion to other unions, 53, 54, 56–57, 60
  financial losses to strikers, 59
  IP lawsuits filed, 79
  Jay (Maine), effect on, 65–66
  leadership during, 54, 56, 58, 63
  media coverage, 54–55, 57, 57*n*, 58–59
  organizing hazards, 51, 54, 56, 59–60
  organizing strategy, 50–52, 54
  prelude to, 51–52, 61
1918 strike, paper workers, 73–74, 171
1919 strike, paper workers, 74–75
1919 strike, steelworkers, 77–78
1921 strike, paper workers
  1987/88 strike compared, 86–87, 89
  chronology, 78–86, 171
  inter-union cooperation, 81
  IP position, 88
  scabs/strikebreakers in the, 80, 82–83, 84, 84*n*
  summarized by UPIU, 89
  support for, 79–80
  violence during, 80–81
1983 strike, steelworkers, 9
1985 strike, United Food and Commercial Workers, 9
1987/88 strike (Local 14, Jay, Maine)
  1921 strike compared, 86–87, 89
  beginnings, 1, 4, 12–14
  cartoons about, 3
  causes underlying, 55
  chronology, 171–176
  contract proposal, 10
  Corporate Campaign implementation, 137–138
  culture of solidarity, 129
  decertification, 5*n*, 176–177
  ended, 14, 63*n*, 138
  failure, reasons for, 9–10, 129
  images characterizing, 4, 6–7
  IP lawsuits filed, 79
  legislation resulting from
    Environmental Control and Improvement Ordinance, 5, 162–163, 173, 175, 177
    permanent replacements outlawed, 8, 8*n*
  media coverage, 3, 3*n*
  political impact
    IP bond refused, 173
    Jesse Jackson supported by, 8*n*

property tax reassessment for IP approved, 4, 176
union members elected to office, 4, 7*n*, 11
pre-strike strategies
IP (International Paper), 1, 8–10
of Local 14, 11–12, 13
UPIU, 10–11, 11*n*, 63*n*
public support, 4, 6–7, 62, 62*n*, 128–129
Rumford strikers relationship, 10, 10*n*, 11
statistics
demonstration participants, 6, 173, 174, 175
media coverage, 3*n*
members walk off job, 172
membership, 5*n*, 8
permanent replacements, 9*n*, 173, 174
rehired workers, 139*n*, 176–177
scabs on job, 173
vote to strike, 172
support functions of Local 14, 4
supported by other unions, 4–5, 5*n*, 10*n*, 13–14, 91, 158
UPIU support/lack of
convention resolution, 175
Getman on, 62–63, 63*n*
by leadership, 62, 128–129
Pine Bluff strike, 13–14
strike benefits, 11, 173
strike called off, 176

A

Adamson Act, 69
AFL (American Federation of Labor)
1935 organizing strategy, 102–103
cartoon, 103
CIO relationship to, 102–104
on Clayton Act, 68
Eagle Lodge charter granted, 26, 170
goals, 155
Great Northern lockout, role in, 41–42
history, 33*n*, 65
NLRA opposed by, 110–111
operating theory, 147
on Sherman Anti-Trust Act, 68
unconstitutional injunctions policy, 154
WWI position, 72
AFL-CIO, 66–67, 141, 157–158, 162
Aid to Families with Dependent Children (AFDC), 97
air traffic controllers strike, 9, 10, 86, 109, 133, 157–158
Allied Industrial workers, 176
Altgeld, John, 22–23
American Federation of Labor (AFL). *See* AFL (American Federation of Labor)
American Newspaper Publishers' Association, 19
American Plan, 78, 91
American Railway Union, 69, 72
American Wire Weavers Protective Association, 171
Androscoggin Mill, 172
Association of Western Pulp and Paper Workers, 77*n*, 141, 171

B

"Bars and Shadows" (Chaplin), 72
BASF Corporation, 117
Bath Iron Works shipyard, 70
Baxter, Governor of Maine, 80
Becker, Craig, 123
Bellows Falls, Vermont, 80
Berlin, IP mill (New Hampshire), 57
Berlin, New Hampshire, 79
The Betrayal of Local 14 (Getman), 62–63
Black, Hugh, 150
Blanchard, I. G., 134
blue-card system of dismissal, 51
Boise Cascade Corporation, 9
Boston Police Patrolmen's Association, 86, 86*n*
Bowdoin Paper Manufacturing Co., 57
boycotts, 20–21, 20*n*, 149–150. *See also* injunctions against labor
Bradshaw, Joe, 13
bread and butter unionism, 147
Building and Construction Trade Unions, 46*n*
Bureau of Industrial and Labor Statistics, 38
Burke, John, 39, 66, 84–85, 99–100, 109, 152
Burns, Matthew, 99–100

Bush, George H., 148, 176

C

Canadian International Paper Company, 107
Capital and Labor polarization, 131
captive audience meetings, 122
Carbonella, August, 57–58, 65–66, 92
Carey, J. T., 42–43, 52, 54, 57
Careyville, IP mill (New York), 57
Carlisle Group, 78–79, 80, 84–85, 171
Carlisle, Fred, 78
Carpenters Union, 77*n*
Carroll, Lewis, 109–110
Carter, Jimmy, 148
Catholic Church, 92
Central Labor Union, Livermore Falls, 67
Champion Paper Company, 11*n*
Chaplin, Ralph, 72
child labor laws, 37, 69
Chisholm neighborhood, 26*n*, 33
Chisholm, Hugh, 17, 17*n*, 26*n*, 35, 59
Chrysler Corporation, 114
CIO (Congress of Industrial Organization), 95, 104–108, 107*n*, 132–134, 171
Clayton Act, 68
Clinton, Bill, 148
closed shop agreements, 45–46, 46*n*, 115
Commission on Industrial Relations, 68
Commonwealth of Massachusetts v Hunt, 149
communists in unions, 115, 171
Conditions of Labor in this Country (Altgeld), 22–23
Congress of Industrial Organization (CIO). *See* CIO (Congress of Industrial Organization)
constitutional amendments
  13th, 21, 91, 153–155, 159–160, 168
  14th, 68, 150, 159–160
  1st, 97, 159–160
Converse Rubber Plant, Maine, 122
Coolidge, Calvin, 86
Corinth, IP mill (New York), 14, 73, 80, 174. *See also* 1910 strike
corporate America
  1930-1980s culture compared, 109
  American Plan devised by, 78, 91
  cartoon, 53
  class division in, 39–40, 68–69, 143
  consitutional protection, 1880s, 150
  courts favor, 114–115
  culture of domination, 10
  forced to recognized unions by legislation, 106
  hegemony accepted by unions, 132–136, 133*n*
  ideology of new world order, 87–88
  merger movement, 17
  Palmer Raids, 88
  power of modern day
    in union organization process, 120–123
  Rockefeller principle regarding unions, 55–56
  role in union organization, 98–102
  war, effects on, 70–71
Corporate Campaign strategy, 11, 13, 63*n*, 136–138, 140, 174–175. *See also* sit-down strikes
corporate capitalism
  Altgeld speech, 22–23
  Burke on, 39
  cartoons on, 18, 34
  court injunctions against labor, 20–21, 68, 77–78
  culture of, 21–22, 55–56, 86–87, 133–135
  debt peonage system, 68–69
  defined, 147
  labor vs. capital, 40, 55, 131–132
  Lincoln on, 40
  trickle down theory, 86, 96–97, 156
  trust/monopoly regulation, 19–20, 68
Corporate Liberalism (Lustig), 20
Cox, James M., 69

D

Dartmouth College, 150
Debs, Eugene Victor, 69, 72. *See also* I.W.W. (Industrial Workers of the World)
democracy
  defined, 142
  industrial, 152–153
Democratic Party and labor, 148, 159
Department of Environmental Protection (U.S.), 174, 175

Department of Labor (U.S.), 67
Depression, the Great, 96–98, 155–156
Deschaine, Lucien, 6
Dinardo, Jimmy, 5*n*
Donne, John, 15
Doughten (organizer), 57
Douglass, Frederick, 127
Downsize This! (Moore), 133

E

Eagle Lodge, 26, 170
Edwards, Phil, 14–15
EMBA (Employees Mutual Benefit Association), 91
environment, IP violations and fines, 139, 162–163, 174, 175, 176
Environmental Control and Improvement Ordinance, Jay, Maine, 5, 162–163, 173, 175, 177
Environmental Protection Agency (EPA), 139
Espionage Act, 72

F

Federal Deposit Insurance Corporation (FDIC), 97
Federal Laborers Protective Union, 26, 27, 170
Federal Wage Board, 73
feudalism, industrial, 68
Fitch, John A., 50–51
Flint sit-down strike (1937), 106, 156
Fort Edward, IP mill (New York), 52, 73
Franklin, IP mill (New Hampshire), 56, 57
Franklin, New Hampshire, 56
Funk, Larry, 12, 172

G

General Motors Corporation, 106, 108, 114, 135–136, 156
Georges, John, 1, 175, 176
Gephart, Richard, 174
Getman, Jack, 62–63, 63*n*
Glenn, Wayne, 1, 11, 13, 62–63, 128–129, 173
Glens Falls, IP mill (New York), 52, 73
glorious labor amendment, 153,-154, 158-159. *See also* constitutional amendments
Gompers, Samuel, 154
good-faith bargaining, 116–117, 155
Goodwin, Lawrence, 161
Graham, Harry, 77*n*, 140–141
Great Northern lockout (1907), 41–42, 46, 49, 50
Great Railroad Strike (1877), 52*n*
Green, William, 100, 110–111

H

Hamilton, IP mill (Ohio), 14, 173, 174
Harding, Warren G., 69
Hayes, Alfred, 167
Haywood, "Big Bill", 89
hegemony, 133*n*
Hill, Joe, 92, 167
"Historical Memory, Class Formation, and Power" (Carbonella), 65–66
Hormel Company, 9
Howard, Charles, 102

I

I.W.W. (Industrial Workers of the World), 69, 72, 89, 123, 133–134
ICC (Interstate Commerce Commission), 19–20
Independent Workers of North American (cement workers), 176
injunctions against labor
  AFL policy on, 154
  cartoon, 20
  Clayton Act, effect on, 68
  corporate capitalism and, 20–21, 68, 77–78
  Norris-LaGuardia Anti-Injunction Act (1932), 49*n*, 106, 154, 155*n*, 156–157
  scabs/strikebreakers protected, 52
  states policies on, 155*n*
  strikes/boycotts/pickets, 46, 106*n*, 115, 136, 147–148, 149–150
  Taft-Hartley Act (1947), 46, 115, 136, 147–148
International Brotherhood of Firemen and Oilers, 5*n*
International Brotherhood of Paper Makers, 171
International Brotherhood of Paper Makers,

Pulp, Sulphite and Paper Mill Workers (IBPMPSPMW), 27, 29
International Brotherhood of Pulp, Sulphite and Paper Mill Workers (IBPSPMW), 27–29, 73–75, 170
International Paper Machine Tenders Union (IPMTU), 26, 170
Interstate Commerce Act (1881), 19
Interstate Commerce Commission (ICC), 19–20
IP (International Paper)
  blue-card system of dismissal, 51
  boycott against, 137–138
  cartoons, 87
  Catholic Church, relationship with, 92
  Employees Mutual Benefit Association (EMBA) formed, 91
  environmental violations and fines, 139, 162–163, 174, 175, 176
  history, 17–19
  holdings, 8, 8*n*, 9, 35*n*
  ideology, 87–88
  locals represented in, 12*n*
  open shops at, 77, 107
  OSHA fines received, 139, 175
  paternalistic attitude of, 32–33, 53, 59
  Peace Accord agreement with UPIU, 138–139
  power of, 19, 35
  profits/losses, 1920-1924, 84
  strategy for organizing workers, 1930s, 107–108
  strike breaking strategies, 51
  War Labor Board award and, 72–75
IPMTU (International Paper Machine Tenders Union), 26, 170

J

Jackson, Jesse, 8*n*, 116, 174
Jay, Maine. *See also* 1987/88 strike
  1910 strike, effect on, 65–66
  chlorine dioxide gas evacuation, 7, 173
  Environmental Control and Improvement Ordinance, 5, 162–163, 173, 175, 177
  IP demands made, 1987/1988, 8
  mill history, 17–18, 22
  post-1910 strike, 65–66
  property tax reassessment for IP approved, 4, 176
"Joe Hill" (Hayes), 167
Jones, Maurice, 50

K

Kellman, Peter
  1987/1988 strike, role in, 1–2
  Carter-Mondale vote directive and, 162
  fired for union organizing, 116
  Republican, election help given, 120
  union activity, 116
  union recognition election participation, 122
Kirkland, Lane, 66–67
Knights of Labor, 33, 33*n*, 38, 123, 133–134

L

labor. *See* organized labor; working class
Labor Board, Roosevelt Administration, 155
Labor Conspiracy Law (1891), 37
Labor Day observance, 25, 33–34, 33*n*, 67, 81, 174
labor movement, modern mature, 133
Labor Party, 159–160
labor vs. capital, 40, 55, 131–132
Laborers Protective Union No. 9,555, 26
Lambertson, George, 86
legislation. *See also* constitutional amendments, United States; NLRA (National Labor Relations Act (1935)); NLRB (National Labor Relations Board); *specific legislation* e.g. Espionage Act
  1987/1988 strike, impact on, 5, 8, 8*n*, 162–163, 173, 175, 177
  captive audience meetings ruled illegal, 122
  child labor, 37, 69
  closed shop agreements outlawed, 115
  by Commission on Industrial Relations, 68
  eight-hour day, 69
  Labor Day observance, 33*n*
  permanent replacements, 8, 8*n*, 37, 114, 157
  trust/monopoly regulation, 19–20, 68

union shop banned, 115
Lewis, John L., 102, 105, 106, 108, 132–133, 147–148
Lincoln, Abraham, 40
Livermore Falls, IP mill (Maine), 17–18, 52, 54, 73
Livermore Falls, Maine
1910 strike, public support for, 54–55
1921 strike, strikebreakers unwelcome in, 80
Labor Day observance, 25, 33–34, 67, 81
mill history, 17–18, 22, 91
National Guard organization in, 52*n*
Local 14, Paperworkers' strike (Jay, Maine). *See* 1987/1988 strike
lockouts, 1, 4, 4*n*, 21, 117, 172. *See also* sit-down strikes
Longshoremen strike (1934), 101–102
Ludlow strike (Colorado), 55
Lustig, Jeffrey, 20
Lynd, Staughton, 61
Lyons, Thomas J., 49

M

Madawaska, Maine, 6
Magnan, T. G., 92, 92*n*
Maine, State of
Bureau of Industrial and Labor Statistics created, 38
IP demands made, 1987/1988, 8
IP fined for environmental violations, 139
Labor Day legislation, 33*n*
political impact of Local 14 strike, 4, 11
pulp and paper industry history, 17–18
strike statistics, 1881-1900, 21
Malin, John, 54, 56, 57
maritime unions strike (1934), 101
"Mark Well That in This Place You Stood" (Edwards), 14–15
mason, fireplace story of, 47
Maurin, Peter, 119
McCarthy, J. J., 168
McFalls, Tommy, 14
McKernan, John R., 55, 175
Meany, George, 148
Menasha, IP mill (Wisconsin), 174
militant, defined, 142
Milton, IP mill (New Hampshire), 57
Mitchell, John, 20–21
Mobile, IP mill (Alabama), 1, 4, 117, 117*n*, 172
monopolies, regulation of, 19–20, 68
Moore, Michael, 133, 134, 135
Moss Point, IP mill (Mississippi), 173
Murphy, Frank, 114
Murray Hall, 91
Murray, W.A. (Bill), 91

N

National Guard, 52, 52*n*
National Industrial Recovery Act (NIRA, 1933), 95, 98
National Labor Relations Act (1935). *See* NLRA (National Labor Relations Act (1935))
National Labor Relations Board. *See* NLRB (National Labor Relations Board)
National Recovery Administration (NRA), 98
Nearing, Scott, 72
New Deal, 155
NIRA (National Industrial Recovery Act (1933)), 95, 98
NLRA (National Labor Relations Act (1935)). *See also* legislation
AFL opposition to, 110
constitutionality of, 105, 114–115, 156
corporate hegemony accepted by, 133
good-faith bargaining defined under, 116–117
purpose of, 113
right to strike under, 117–118
union growth and, 95
union purpose defined under, 140
yellow dog contracts under, 49*n*
NLRB (National Labor Relations Board). *See also* legislation
bargaining units prior to, 38, 41
bargaining units restrictions of, 46
legal process outlined, 118
organizing units, restrictions for, 46
pool process ruled illegal, 119

purpose of, 106, 116, 124, 155
union certification election process, 120–121, 160
utilization by corporations, 114
NLRB v Jones & Laughlin Steel Corp, 114, 156
NLRB v Mackay Radio, 114, 157
NLRB v Virginia Electric & Power Co., 115, 122, 157
Norris-LaGuardia Anti-Injunction Act (1932), 49*n*, 106, 154, 155*n*, 156–157
North Walpole, New Hampshire, 80
NRA (National Recovery Administration), 98

O

Office and Professional Employees International Union (OPEIU), 5*n*
Oil, Chemical and Atomic Workers Union (OCAW), 28*n*, 117, 177
Olney, Richard, 19–20
O'Neil, Paul, 8*n*, 172
open shops, 49, 49*n*, 77, 91, 107
organized labor. *See also* working class
1830s, 152
1884-1908, 33, 33*n*, 39, 43–47, 49, 49*n*
1921-1926, violence used by, 89
1940s, 114–115
1970s, 116
call to action, 168–169
culture of hope in, 168–169
union history, 149–151
organized labor, 1909-1920. *See also* 1910 strike
Catholic Church, relationship with, 92
culture of, 58, 61, 65–66
force used to subdue, 52, 52*n*, 54–56
growth of, 65
Labor subservient to Capital, 131–132
political involvement, 67–69
power of, 58, 71, 120
Rockefeller principle regarding, 55–56
strike breaking strategies by corporations, 59–60
trust/monopoly regulation of, 68
violence against, 92, 167
vision of, 124, 153
WWI, impacts on, 67–69
organized labor, 1930s
AFL organizing strategy, 102–103
corporate ownership of, 99–102, 107
cultural change in, 96
Labor subservient to Capital, 131–132
legislation weakening, 114–115
member-based to employer-based, 98, 99–102
organizing process, 100–101
power of, 101–102, 106–108, 114, 120, 156
regulation of, 97, 98
resurgence/growth, 95–97, 99, 102, 107–108
Roosevelt administration and, 96–99, 155–156
Uniform Labor Agreement (ULA) established, 100
violence against, 101
organized labor, modern day
bargaining unit size and contract negotiations, 118–119
contract negotiations limitation, 118–119
Corporate Campaign implementation, 136–138
corporate hegemony accepted by, 132–136, 133*n*
culture of, 61, 120
grievance procedure, 160–161
media coverage, lack of, 66
power of, 46–47, 118–119
problems inherent to, 61–63, 62*n*
strategy for organizing workers, 109, 148–149
union organization process, 120–122
union shop agreements, 46*n*
vision of, 123–124, 146–148, 152–155
OSHA, 139
Otis mill, 12, 17–18, 17*n*, 33, 172

P

PACE (Paper, Allied-Industrial, Chemical and Energy International Union), 28*n*, 113*t*, 139, 141, 177
Palmer Raids, 88
paper bag makers strike, 1906, 35

Paper Makers Union (PMU). *See also* 1910 strike; Paper Makers-Pulp & Sulphite Unions, UPIU (United Paperworks International Union)
- 1903 strike, 168
- 1907 Great Northern lockout, 41–42
- 1908 strike, 42–47, 50
- 1921 strike, 79, 84–86
- 1964 split, Association of Western Pulp and Paper Workers formed, 77*n*, 141, 171
- chronology, 170–176
- membership statistics, 25*t*, 29, 65*t*, 77*t*, 95*t*, 113*t*, 171
- power of, 49
- Uniform Labor Agreement (ULA) and, 100
- union shop agreement with IP, 108

Paper Makers Union Local #11, 66
Paper Makers-Pulp & Sulphite Unions. *See also* 1910 strike
- 1907 Great Northern lockout problems, 41–42
- 1908 strike conflicts, 42–47, 49
- 1909 truce, 49–50
- 1910 strike outcomes, 66
- 1921 solidarity break, 79, 84–85
- 1964 split, 77*n*
- 1972 merger, 172
- inter-union cooperation, 66, 81
- inter-union problems, 27–30, 50

Paper Mill Workers Union, 108
The Paper Rebellion (Graham), 77*n*, 140–141
Paper Trust, 19, 35
Paper, Allied-Industrial, Chemical and Energy International Union (PACE). *See* PACE (Paper, Allied-Industrial, Chemical and Energy International Union)
Peace Accord agreement, 138–139
perfecting the organization (concept), 52*n*, 56, 58, 110, 129
Perfecting the Organization (process), 120
Phelps Dodge Company, 9
Philadelphia Cordwainers Case, 149
pickets, 20–21, 52, 104, 174. *See also* injunctions against labor
Pine Bluff, IP mill (Arkansas), 10*n*, 13–14, 174
Pittston Coal Company, 127–128
political action vs. politics, 162–164
political elections
- CIO involvement in, 104, 106
- Jesse Jackson supported by unions, 8*n*
- Roosevelt presidency supported by labor, 156
- Wilson presidency supported by labor, 67–69

politics. *See also specific parties*, e.g. Socialist Party
- political action vs., 162–164
- working class involvement in, 39–40, 88–89

The Populist Moment (Goodwin), 161
Poulin, Mickey, 107
poverty, cycle of, 118
"The Preacher and the Slave" (Hill), 93
Project Productivity, 8
Prudent Man rule, 136
Pullman strike of 1894, 69, 72
Pulp & Sulphite Union. *See also* 1910 strike; Paper Makers-Pulp & Sulphite Unions; UPIU (United Paperworks International Union)
- 1907 Great Northern lockout, 41–42, 46, 50
- 1908 strike, 42–47, 50
- 1921 strike, 79, 84–86
- 1964 split, Association of Western Pulp and Paper Workers formed, 77*n*, 141, 171
- chronology, 170–176
- class division in, 123–124
- history, 12, 27–29, 33
- Local 299, 115
- membership statistics, 25*t*, 29, 65*t*, 77*t*, 95*t*, 108, 113*t*, 170–172
- Uniform Labor Agreement (ULA) and, 100
- union shop agreement, 1930's, 108

Q

Quin, Mike, 101

R

radical, defined, 142
Railway Labor Act (1926), 106
Reagan, Ronald, 9, 10, 86, 86*n*, 109, 133, 148, 157
replacement worker, 29*n*
replacements, permanent. *See also* scabs; strikers
  1908 strike, 52
  1987/88 strike, 9*n*, 14, 173, 174
  Boise Cascade strike, 9
  legislation regarding, 8, 8*n*, 37, 114, 157
  Mobile, IP mill (Alabama), 117
  NLRA regulation on, 117–118
replacements, temporary, 117
Republican Party and labor, 148
Rockefeller, John D., 55–56
Roderick, Royal, 12
Rogers, Ray, 63*n*, 136–138, 137*n*, 174
Roosevelt administration (F.D.R.), 96–98, 104, 106, 109, 132, 155
Roosevelt, Theodore, 69
Rumford mill (Maine), 9, 10*n*, 11, 35, 59, 73, 80
Rumford, Maine, 1910 strike support, 54, 59

S

Sacco, Nicola, 89
Santa Clara, 150
scabs. *See also* replacements, permanent
  1907 Great Northern lockout, 50
  1908 strike, 29–30, 49, 52
  1910 strike, 52, 54, 57
  1921 strike, 82–83, 84, 84*n*
  1987/88 strike, 173
  cartoon, 83
  defined, 4*n*, 29*n*, 30
  super-scab apology, 7
Schneider, George, 54, 56, 59, 100
Seamen's Act, 68–69
Seitz, Mr., 19, 35
Service Employees International Union, 123
Sherman Act (1890), 19–20, 68
shoemakers union, 149
sit-down strike movement, 156
sit-down strikes, 104, 106, 114, 156. *See also* Corporate Campaign strategy; strikes
slave labor, 149–151
Slave Labor Act. *See* Taft-Hartley Act (1947)
Social Security, 97, 132*n*
social welfare programs, 38–39, 67, 78, 91, 97
Socialist Party, 69–70, 89, 170
solidarity, 62–63, 79, 84–85, 129, 152, 158–159
Solidarity Day march, 157–158
"Solidarity Forever" (Chaplin), 72
Solin, IP mill (Maine), 56
Southern Kraft Paper Company, 108. *See also* IP (International Paper)
St. Regis Paper Company, 78
Steel Workers Organizing Committee, 108
Steelworkers strike, 9, 77–78
strategy, defined, 145
strikebreakers, 52, 80. *See also* scabs
strikers, 157. *See also* replacements, permanent
  in 1987/88 strike, 5, 7
  women as, 21, 21*p*, 35
strikes. *See also* Corporate Campaign strategy; sit-down strikes
  history, 1881-1900, 21–22
  injunctions against, 20–21, 46, 106, 106*n*, 115, 136, 147–150
  process for calling/ending, 13*n*
  right to under NLRA, 117–118
  statistics on participation in, 21, 72*n*, 73*n*
Sullivan, Herbert, 100
Supreme Court, threat to stack, 156

T

tactic, defined, 145–146
Taft, William Howard, 69, 71
Taft-Hartley Act (1947), 46, 115, 136, 147–148
Teamsters boycott (1934), 101
Teamsters Union, 141*n*
theory, defined, 145–146
Three Rivers Mill (Quebec), 84, 85–86
Through the Looking Glass (Carroll), 109–110
Ticonderoga, IP mill (New York), 57, 73, 175, 176

Tierney, James E., 175
trade secrets of workers, 22, 45
Treaty of Peace, 74
Trethewey, B. H., 29
trickle-down/trickle-up theory, 86, 96–97, 156
trusts, 19–20, 19*n*, 68

U

UAW (United Auto Workers). *See* United Auto Workers (UAW)
UBPM (United Brotherhood of Paper Makers), 26, 27–29, 170
UMWA (United Mine Workers of America)
  cartoon, 103, 105
  CIO victories compared, 132
  culture of solidarity, 129
  election process, 141*n*
  expelled from AFL, 103
  history/tradition of, 67, 89, 127
  leadership, 129
  organizing strategy, 102, 108, 108*n*
  Pittston Coal Company defeated, 127–128
  reorganization, 1980s, 127–128
Unemployment Insurance, 97
Uniform Labor Agreement (ULA), 100
union, purpose of, 67, 140. *See also* organized labor
union bargaining units, 38, 41–42, 46
union certification election process, 120–122, 157, 160
union hall, the, 7, 66, 91
union leadership
  1910 strike, 54, 56, 58, 63
  1987/88 strike, 62
  affiliation with the corporation, 98, 137–138, 140, 142
  at CIO, 104
  election process, 140, 140*n*, 141–142, 153
  Kirkland presidency, 66–67
  Moore on, 133, 135
  relationship to active membership, 140–141
  solidarity, method for achieving, 158–159
  at UMWA, 129
  at UPIU, 62–63, 62*n*, 128–129
union locals, election process, 141–142
union membership
  1908 strike, before/after, 45, 49
  1910 strike, after, 65–66
  benefits of, 29, 38–39, 47, 66–67, 140–142
  cultural components, 47, 58, 151–152, 168
  relationship to leadership, 140–142
  union shop agreements and, 46–47
  WWI, impacts on, 71
  yellow dog contracts and, 49*n*
union membership statistics. *See also specific unions*, e.g. Paper Makers Union
  1886, 33*n*
  1900-1908, 25*t*, 168, 170
  1900-WWI, 145
  1908 strike, before/after, 29
  1909-1920, 65*t*, 170
  1920 vs. 1932, 145
  1920-1923 change, 78
  1921-1935, 77*t*
  1930s, 145, 171
  1936-1961, 95*t*
  1948, 171
  1950s, 171
  1962-2001, 113*t*
  1964, 171
  1975, 145
  as percentage of workforce, 145, 148
  present day, 123, 131, 148
  in Republican vs. Democratic presidencies, 148
union organization, process overview, 120–123
union privilege card, 66–67
union publications, 39, 66
union recognition elections, 38, 116, 120–122
union shop, 46, 115
union-free environment (open shop). *See* open shops
United Auto Workers (UAW), 106, 114, 123, 133, 135, 156
United Brotherhood of Paper Makers

(UBPM), 26, 27–29, 170
United Electrical Workers, 141*n*
United Food and Commercial Workers, 9
United Mine Workers of America (UMWA). *See* UMWA (United Mine Workers of America)
United Paper Workers of America, 171
United Wallpaper Craftsmen and Workers of North America, 171
UPIU (United Paperworkers International Union). *See also* Paper Makers Union; Pulp & Sulphite Union
- 1987/88 strike involvement, 10–11, 11*n*, 13–14, 63*n*
- Association of Western Pulp and Paper Workers affiliation, 77*n*
- Boise Cascade strike, 9
- Corporate Campaign implementation, 137–138, 137*n*, 140
- culture of defeat, 139
- hierarchical structure, 141
- history, 27–29, 28*n*
- leadership, 62–63, 62*n*, 128–129, 140
- membership statistics, 8, 113*t*, 172, 176, 177
- Peace Accord agreement with IP, 138–139
- power of, 138–139
- strike process, calling/ending authority, 13*n*

V

Vanzetti, Bartolomeo, 89
vision, of organized labor, 146–148

W

Walsh, Frank, 68, 71
war. *See also specific wars*
- benefits to the corporation, 70–71
- strike compared to, 14–15

War Labor Board, 68, 71–74, 153, 171
"We mean to make things over" (Blanchard), 134
wealth, cycle of, 118
welfare for workers. *See* social welfare programs
"When the workers" (Maurin), 119
whipsaw process, 71
Wilson, Charles E., 114
Wilson, William, 67
Wilson, Woodrow, 67, 68–69, 71, 153, 156
women workers, 21, 21*p*, 35, 70, 70*n*, 73
work rules, 117
worker wages
- 1900-1909, 35, 41, 47, 49
- 1910-1919, 73–74
- 1920s, 84
- 1930s, 100, 108
- loss, in real dollars, 118, 123, 143

Workers' Compensation, 8
workers rights
- eight-hour day, 40, 47, 60, 69, 73, 74
- forty-hour week, 108
- shutdown holidays, 12, 27, 33, 41, 49, 55, 60, 73, 74, 100

working class vision, 149–152
working class, 1881-1908
- class consciousness in, 33
- class division in, 39–40
- culture of, 21–23, 33, 39, 168
- mobility of, 45–47
- political involvement of, 43
- power of, 22, 45–47

working class, 1909-1921
- class consciousness in, 58
- class division in, 68–69
- culture of, 51, 57–58, 63, 65–66, 120, 124
- political involvement of, 88–89
- social welfare programs for, 78
- WWI, impacts on, 70, 70*n*

working class, 1920s
- Catholic Church and, 92
- social welfare programs for, 91

working class, 1930s
- class consciousness in, 104
- class division in, 132
- culture of unity, 120
- social welfare programs for, 97
- unemployment statistics (1933), 97
- union negotiations, involvement in, 100
- workers' rights enforced, 106, 109

working class, modern day
- class division in, 123–124, 143
- collective rights for, 160–161, 164

freedoms of, 122
justice available to, 116, 118
Labor Party for, 159–160
rights of, 148
solidarity among, 62, 158–159
working conditions
child labor, 37, 69
colonial times, 149–150
described, 1895, 31
industrial accidents, 31–32
internationally legislated, 159
IP statement regarding, 53
at Otis mill, 17–18
yellow dog contracts required, 49*n*
World War I, 67–72, 70*n*, 152–153, 156
World War II, 70

Y

yellow dog contracts, 49, 49*n*, 154–155
Young, Boyd, 139

Z

Zieger, Robert H., 95, 107–108